THE
PERFUME
LOVER

DENYSE
BEAULIEU

A PERSONAL HISTORY OF SCENT

ST. MARTIN'S PRESS
NEW YORK

Some characters described in this book are composites,
and certain events have been compressed. The names and identifying
characteristics of the individuals involved in the author's personal life
who appear in these pages have been changed.

ISBN 978-1-250-02501-2 (hardcover)
ISBN 978-1-250-02502-9 (e-book)

St. Martin's Press books may be purchased for educational, business,
or promotional use. For information on bulk purchases, please contact
Macmillan Corporate and Premium Sales Department at 1-800-221-7945
extension 5442 or write specialmarkets@macmillan.com.

First published in Great Britain by Collins,
an imprint of HarperCollins*Publishers*

First U.S. Edition: March 2013

10 9 8 7 6 5 4 3 2 1

Prologue

I am sitting in the perfumer's lab, taking delicate cat-like sniffs of a slender blotting-paper strip dipped in orange blossom absolute. Though the flower has long been worn by brides as a symbol of purity, what I smell on this strip drops more than a hint of the earthier proceedings of the wedding night. In fact, the candid orange blossom, just like her sisters jasmine, gardenia and tuberose, hides quite a whiff of sex under her tiny skirts. No wonder the aphrodisiac essences of white flowers have always held pride of place in the arsenal of French perfumers. They are subtle reminders of the beast lurking within the beauty. Perfume is meant to go on a body; animal notes are what weld the suave scent of floral flesh to ours. Isn't it a little hot in here?

While the sweet, narcotic orange blossom absolute unfolds its facets, the perfumer talks me through its more unexpected aspects. His words are like conjuring tricks: as he speaks, a snapped peapod, a whiff of hot tar just before a storm, a fatty smear of wax, a dusting of honeyed pollen spring from the strip. Orange blossom absolute is animal, vegetable and mineral all at once; it doesn't quite smell like the flower it is extracted from. Still, there is enough of its ghost hovering over the blotter to tug at a distant memory. I close my eyes and let the fragrance seep into my mind, until it overwhelms all the other smells in the lab, bringing me back to the first time I ever breathed it in ...

It had another, more poetic name for me then, bequeathed by the Moors to the Spanish language: *azahar*.

I am in Seville, standing under a bitter orange tree in full bloom in the arms of Román, the black-clad Spanish boy who is not yet my lover. Since sundown, we've been watching the religious brotherhoods in their pointed caps and habits thread their way across the old Moorish town in the wake of gilded wood floats bearing statues of Christ and the Virgin Mary. This is the Madrugada, *the longest night of Holy Week, and the whole city has poured into the streets: the processions will go on until the dawn sky is streaked with hunting swallows. In the tiny whitewashed plaza in front of the church, wafts of lavender cologne rise from the tightly pressed bodies. As altar boys swing their censers, throat-stinging clouds of sizzling resins – humanity's millennia-old message to the gods – cut through the fatty honeyed smell of the penitents' beeswax candles.*

Under the silver-embroidered velvet of her dais, the Madonna, crystal tears on her cheek, tilts her head towards the spicy white lilies and carnations tumbling from her float. She is being carried into the golden whorls of a baroque chapel, smoothly manoeuvred in and out, in and out, in and out – they say the bearers get erections as they do this – while Román's hand runs down my black lace shift and up my thigh to tangle with my garter-belt straps. His breath on my neck smells of blond tobacco and the manzanilla wine we've been drinking all night – here in Seville, Holy Week is a pagan celebration: resurrection is a foregone conclusion and there is no need to mourn or repent. As the crowd shifts to catch a last sight of the float before the chapel doors shut behind it, the church exhales a cold old-stone gust. I am in the pulsing, molten-gold heart of Seville, thrust into her fragrant flesh, and there is no need for Román to take me to bed at dawn: he's already given me the night.

Bertrand Duchaufour is leaning towards me with the rapt gaze of a child listening to a fairy tale. He removes his glasses and wipes them, nodding, before his lips curl into a boyish grin.

'All those smells … it's all there! Now *that* would make a very good perfume.'

I've just been leading one of the world's best perfumers by the nose.

1

I'd have never imagined that some day I'd be telling Bertrand
Duchaufour about my nights in Seville. When we first met in
a radio studio in May 2008, I hadn't even liked him much.

I'd only been writing about fragrance in earnest for a year at
that point and I'd been very much looking forward to meeting
Duchaufour. His off-beat, deeply personal compositions for
edgy fashion labels like Comme des Garçons or the pioneer
niche house L'Artisan Parfumeur had earned him star status
among perfume aficionados as well as a reputation for artistic
integrity. He was one of the people who'd eased fragrance out of
its traditional set of references as projections of feminine or
masculine personae: many of his compositions were olfactory
sketches capturing the spirit of the places to which he'd travelled:
Sienna in winter; a seduction ritual in Mali; a Buddhist temple
in Bhutan; the Panamanian rainforest; a church in Avignon …

With his rectangular glasses, shaven pate and forthright
demeanour, the forty-something perfumer certainly looked more
like one of my artist friends than the sibylline master of a secret
craft, and I was sure we'd get on fine. I was mistaken. Throughout
the broadcast, Duchaufour was gruff and snappy. The host's

faux-naïve questions seemed to irk him; he let on that self-styled critics like myself or our fellow guest, the perfume historian Octavian Coifan, had better leave the thinking to the pros. It turned out he had good reason to be annoyed. He'd been led to believe he'd been invited to speak about his work. Minutes before going on the air, he was told that the topics of the day would be the high price of perfume and the absence of proper fragrance reviews. I understood why he was disgruntled and respected the fact that he didn't try to ingratiate himself with us or the public, but I was disappointed just the same. Clearly, we weren't going to be buddies. Still, the man made great perfumes and that was all that mattered. I didn't have to like him personally to appreciate his work, and I certainly didn't need him to like *me*.

So when I spotted him the following November at the raw materials exhibition organized by the Société Française des Parfumeurs where I had just been accepted as a member, I wondered whether I should even bother to say hello. But since we'd met, I'd managed to slip a foot in the door of a few labs: some of his colleagues seemed to think I was worth talking to. And more significantly, I'd fallen in love with his recent work. I felt he was shifting towards a more sensuous style I could actually connect with, and I'd just written a review of his latest fragrance, Al Oudh, an ode to Arabian perfumery pungent with sexy animal notes. 'I love it when a man plays that kind of dirty trick on me,' I'd concluded teasingly. It hadn't occurred to me that he would ever read those words.

Duchaufour recognized me as I walked by: I *was* somewhat conspicuous with my apple-green coat and silver bob. Much to my surprise, he grinned and kissed me on both cheeks before congratulating me on the accuracy of my review. I hadn't written it to please him, but I wasn't about to let such an opportunity pass me by, so I instantly improvised a white lie. I was teaching a perfume appreciation course at the London College of Fashion in a month's time, I told him (which was the truth), and I intended to discuss his work (also the truth, as of one second ago). It was the very first time I was to teach the course, which I'd been offered on the strength of my writing and talks I'd given to students in Paris. At first, I'd felt pretty confident I could swing it, but as the time to shut myself inside a classroom for three days with fifteen eager perfume lovers drew nearer, I was feeling a little jittery. I kept that to myself, but I did tell Duchaufour I'd appreciate his input (if anything, I'm a quick learner). Much to my relief, he nodded, still grinning:

'You're ready to learn more. Come over to the lab whenever you want!'

I wrote to him the very next day to take him up on his offer. After we'd exchanged a couple of emails, he suggested the use of the more familiar French form of address, *tu*, slyly adding, 'It sounds more serious.' So here I am, six months after our first, inauspicious encounter, perched on a chair in his tiny lab above L'Artisan Parfumeur's flagship store and very ready indeed to learn more. The staid sandstone façade of the Louvre looms across the street; the searchlights of a *bateau-mouche* sweep from the Seine over the Pont des Arts. A three-tiered array of neatly labelled phials, each containing one of the hundreds of raw materials of the perfumer's palette, throws amber, topaz and emerald glints under the desk lamp. A paper sheet lies next to a small electronic scale: the forty handwritten lines of the formula he is currently working on. At his feet, three shopping bags bulge with dozens of discarded phials – less than one per cent of his work, he says, ever makes it to the shop shelves. I've just tucked into my handbag a tiny atomizer of a scent of his due to be launched next spring, a tuberose perfume whose working title is 'Belle de Nuit'. Though it was conceived long before we met, it feels like a sign: the tuberose resonates deeply with my life and loves, though he can't possibly know it …

The churlish man who'd snubbed me has turned out to be warm, friendly and almost disconcertingly straight-talking; an intensely focused listener given to boyish bursts of enthusiasm.

About the story of Seville I've just told him, for instance. He loves it, he says it would make a good perfume, but I don't know him well enough to ascertain whether he's the type to follow through or if this is just a perfumer's version of a chat-up line. And certainly not well enough to ask him straight out if he'll do it. Why would he bother with what must be, for him, just one of a hundred different ideas? On the other hand, why wouldn't he? His ideas *do* have to come from somewhere. I didn't tell him my story because I thought it would inspire him. It just came up as we were swapping tales of far-flung journeys. But now this idea is hovering between us and I realize I want this perfume to happen more than I've wanted anything in a very long time. Why couldn't I be a perfumer's muse? I've come such a long way in the realm of scent, Bertrand, you couldn't ever know … In fact, I was never really meant to poke my nose into it.

2

My father couldn't stand perfume. My mother only found out after they got married: she'd never been able to afford fragrance until then. Her first bottle of Chanel N°5 exited their flat soon after she'd brought it home. After that, she never wafted anything stronger than our doctor-recommended Ivory Soap.

As an added excuse for the ban, I was diagnosed with a slew of allergies: cat dander, dust mites, assorted pollens and ... fragrance. But though I grew up on a continent where allergies are a way of life, my parents decided I would continue to be exposed to allergens until I built up a resistance to them. Our tabby still curled up wherever she pleased, tucking her squirming litters in my dad's king-sized Kleenex boxes. Perfume was the only item on the doctor's list to remain strictly *verboten*. I was six years old, and didn't care. A childhood deprived of N°5 doesn't quite register as abusive, and as long as we kept the cat, I was content.

So I don't come equipped with the perfume lover's standard-issue Proustian paraphernalia. No fond memories of dabbing myself with the crystal stopper of Mommy's precious Miss Dior or kissing Daddy after he'd slapped his cheeks with Old Spice;

no tales of lovingly collected perfume miniatures. Nothing but a trail of crumpled-up tissues strewn in an antihistamine-induced daze. I sometimes sneezed so hard I could've propelled myself downtown had I been on roller skates, and my allergies pretty much left me nose-blind. Anyway, in the Plastic Sixties, the suburbs of Montreal didn't give off much more than the smell of burning leaves in fall and fresh-mown lawns in spring and summer – but at the first buzz of a lawnmower, I was sent indoors with my microscope, books and Barbie dolls, lest my bronchial tubes start shutting down. Scented memories of childhood? Access denied.

Bertrand Duchaufour frowns.

'Really? Nothing?'

Could he help me kick-start my memory again? After all, it worked with the orange blossom last week … I've just dropped by to take my second informal lesson and we've been zipping through some of the raw materials he's used in recent scents. I'm on familiar ground until he mentions something called yara-yara. Yara-yara? Sounds like I ought to start swaying my hips to it.

Out comes a blotter and into the phial it goes. He waves it under my nose. Diluted, yara-yara smells of orange blossom. At this concentration: penetrating, narcotic, with a side helping of mothballs. And somehow familiar.

'This reminds me of the wintergreen top notes in tuberose.'

'I wouldn't say so … Here, I'll show you the difference.'

He snatches another phial from the refrigerator and we repeat the blotter ceremony.

Ah … This I know. In fact, I feel like I've always known it …

As it turns out, I *do* have scented memories. But mine come courtesy of Big Pharma. Not only because antihistamines allowed me actually to breathe through the nose every once in a while, but because my father was a pharmacologist.

My visits to his lab as a little girl were thrilling and slightly scary events. The emergency chemical showers in the hallways hinted at the permanent risk of toxic splashes and horrible burns. My dad obviously did a dangerous, heroic job and his lab was one of the most glamorous places in the world to me. It was a smelly place too, permeated with the reek of disinfectants, the salty musky odour of guinea pigs and horsy effluvia – the lab manufactured an oestrogen extracted from the urine of pregnant mares, and we had to drive by the stables to park at the back of the building where my father worked.

Is that why I'm so happy to hang around perfume labs? And why I tend to be drawn to the weird notes that make people blurt out 'Yuck, why would anybody want to smell of *that*?' The animal and medicinal zones of the olfactory map attract me. For instance, the nostril-searing aroma of Antiphlogistine, an analgesic pomade found in every Canadian household since 1919, which was one of the few smells strong enough to burrow its way through my stuffy sinuses …

My growing pains may well have been one of the reasons why I fell in love with the fragrance aptly named Tubéreuse Criminelle for the way it assaults the nostrils when first applied. I am so addicted to it that, some days, I'll spray myself time and again just to catch its venomous minty-camphoraceous blast before the scent subsides in creamy floral headiness. As it turns out, it's also a blast from the past since tuberose and Antiphlogistine have one thing in common: the ice-green burn of a molecule called methyl salicylate.

I know. Those chemical names are a bitch. You don't need to learn them to appreciate perfume, but it helps if you want to make sense of it. Odorant molecules are the building blocks of perfumery. A natural essence may contain hundreds of them. The ones that contribute the most to its smell can be isolated and synthesized. Each will yield a distinct facet of the original essence – of several, in fact, since the same odorants keep popping up all over nature. These molecules can then be assembled to conjure an olfactory illusion in a few broad strokes: when your fragrance has a jasmine note, there's a good chance it comes from a combination of the chemicals naturally found in jasmine rather than its actual essence, of which there might only be a few drops just so the advertising copy doesn't lie. Taken separately, those molecules won't actually smell of jasmine. But if you blend benzyl acetate (floral, apple, banana, nail polish), hedione (green, citrusy, airy), jasmolactone (buttery, fruity, coconut, peach) and indole (mothballs, tooth decay), you'll get a decent impression of the flower that's a lot cheaper to produce than the real stuff. Aromatic materials, natural and synthetic, can also be combined to reproduce the scent of a flower whose essence can't be extracted, such as gardenia, lily-of-the-valley or lilac. And that's not even mentioning the synthetics that smell entirely manmade … You didn't think the musk in your Narciso Rodriguez for Her grew on bushes, did you?

Despite the breathless sales pitches, if fifteen per cent of what's in your bottle comes from a thing that was alive at some point, you're doing well. Any more would cut into profit margins. Natural materials are not always more expensive than synthetics, but they're harder to source. A drought, a flood, a war will make prices shoot up. Crops don't smell exactly the same from one year to another, so that you may have to mix essences from different sources to achieve the same effect in every batch of perfume, a practice called the *communelle*. With synthetics, on the other hand, you can produce batch after batch without worrying about Nature's tantrums or geopolitical flare-ups.

It's not just a matter of price or convenience. Glamorous, exotic and irreplaceable as natural essences may be, it is to synthetics that we owe modern perfumery, and many of the greatest breakthroughs came about as perfumers learned to use them. Synthetics allowed perfumers to structure their compositions by strengthening the relevant facets of natural materials; to conjure the desired effect with a few notes rather than having to draft the entire orchestra of the natural essence; to produce entirely new perfumes. In fact, without synthetics, perfumery would exist neither as an industry, nor as an art.

When Gabrielle Chanel asked Ernest Beaux to come up with a fragrance for her couture house, she told him two things, or so the legend goes (Chanel was not above retro-engineering her life story when it suited her purposes). First: 'A woman must smell like a woman, not like a rose,' a dig at her arch-rival of the time, the couturier Paul Poiret, whose logo was a rose and whose perfume line, the first ever to be launched by a couturier, was named Les Parfums de Rosine after one of his daughters. Chanel, a keen follower of the avant-garde, thought the figurative school of perfumery – 'a rose is a rose is a rose' – was as hopelessly outdated as the plumed and flowered hats she'd replaced with straw boaters. She didn't care much either for the vampy scents her contemporaries doused themselves in. Her perfume, she decreed, should smell as clean as the soap-scrubbed skin of her friend, the famous courtesan Émilienne d'Alençon. But on the eve of the 20s, luxury perfumes still relied extensively on natural essences, many of which ended up leaving a rather rancid odour on the skin because flower extracts were often obtained by spreading the blossoms on mesh frames smeared with fat, a method called enfleurage. Fortunately, Ernest Beaux had a trick up his sleeve, a synthetic material he'd already been playing around with …

Since the late 19th century, organic chemistry had made giant steps, providing perfumers with synthetic materials that were stronger, cheaper, more stable and more readily available than natural materials (availability would particularly become an issue during World War I and its aftermath). These new synthetics allowed perfumers to forego references to nature. Just as the invention of the paint tube had allowed the Impressionists to set up their easels wherever they pleased to catch variations in light, or as the advent of photography had freed painters from naturalistic representation, technical progress fuelled new paradigms in perfumery. But synthetics were often harsh-smelling and only the best perfumers had the skill, the inspiration or the audacity to blend them into a high-quality product. Ernest Beaux was such a man.

On their own, aliphatic aldehydes give off a not particularly pleasant smell of citrus oil mixed with snuffed candles and hot iron on clean linen. They were mainly used in the synthetic versions of natural essences like rose because they had the property of boosting smells, which meant they were mostly found in cheaper products, or in small amounts in finer fragrances to produce fresh, clean, soapy notes – for instance, in Floramye by L.T. Piver, which happened to be a favourite of Mademoiselle Chanel's ... Beaux's genius was to use aldehydes both for their booster effect *and* for their specific smell; to blend them with the noblest, most expensive raw materials; to figure out that they could produce just the required freshness by lifting the heavy scent of the oils and counteracting their rancidness. In the formula of what was to become N°5, he injected an unprecedented one per cent. Later on, he would write that he had been inspired by the icy smell of lakes and rivers above the Arctic Circle.

Did he produce the formula to Gabrielle Chanel's specifications? Or had he already composed it for the company he was working for when Chanel sought him out? Beaux worked for Rallet, a supplier to the Tsar's court, which had been forced by the Bolshevik revolution to move its operations to the South of France. Some industry old-timers claim that as Rallet didn't have the means to exploit its products on a large enough scale (most of its assets had been abandoned in Russia), it decided to offer the formula of its Rallet N°1 to Chanel ...

Whatever the truth of the story, the official legend is as much a part of the perfume as its actual substance: a masterpiece in its own right. But it was that very legend that clouded my perception of N°5 until I stumbled on a pristine, sealed 30s bottle that ripped the veil. Then, at last, I understood its radiant, abstract beauty because the raw materials in it were much closer to the ones Ernest Beaux had used to compose it; their subtle differences were sufficient to jar me out of the cliché that N°5 had become. If I love N°5 now, it is because of the sheer artistry of it, and I discovered that artistry not because it awoke fond memories, since I had none, but for the opposite reason: because of its strangeness. That strangeness is the very reason that led me to the once-forbidden realm of perfumery.

Some people have a signature fragrance that expresses their identity and signals their presence; its wake is an invisible country of which their body is the capital. I'm not one of those people.

I am a scent slut.

'To seduce' means 'to lead astray', off familiar paths and into thrillingly uncharted lands. To me, perfume is not a weapon of seduction but rather a shape-shifting seducer. I have been exploring the world of fragrance in the same way, and for the same reasons, that I've travelled erotic territories, spurred on by intellectual curiosity, sensuous appetites and the need to experiment with the full range of identities I could take on. And just as my experimental bent has driven me to different men, situations and scenarios to find out what I would learn through them, it has led me to different scents.

Perfume is to smells what eroticism is to sex: an aesthetic, cultural, emotional elaboration of the raw materials provided by nature. And thus perfumery, like love, requires technical skills and some knowledge of black magic. Both can be arts, though neither is recognized as such. And I've been studying both in the capital of love and luxury, Paris, where I settled half a lifetime ago. It is in Paris that I learned about *l'amour*; in Paris that I stepped through the looking glass into the realm of scent. I've had good teachers: discussing the delights of the flesh as passionately and learnedly as you would speak about art or literature is one of the favourite pastimes in my adopted country. Here, pleasure is intensified by delving into its nuances. By putting words to it. *La volupté* is taken very seriously indeed, a worthy subject for philosophizing in the boudoir. And so I've come to think of perfumes as my French lovers – a way for gifted artists to seduce me, *parlez-moi-d'amour*-me and reflect the many facets of my soul in eerily perceptive ways …

Blame Yves Saint Laurent and the Frenchwoman who revealed his existence to me via the first drop of perfume ever to touch my skin.

3

I was eleven when I decided I'd be French one day. Not only French, but Parisian. And not only Parisian, but *Left Bank* Parisian: glamorous, intellectual and bohemian.

When she moved into the house next door with a German engineer husband, Geneviève didn't quite replace *The Avengers'* Mrs Emma Peel as my feminine ideal. Despite her closetful of clothes with Paris labels and her collection of French glossy magazines, Geneviève was still a housewife stuck in the suburbs of Montreal and even at eleven I knew I'd never be that. But when she opened that closet and those magazines, she drew me into a world where she herself had probably never lived. The world I live in now.

The scientific community is nothing if not international, but though my parents' cocktail parties could have been local branch meetings of UNESCO, I'd never met a French person before. Next to Geneviève's, my Quebec accent sounded distressingly rustic and I soon applied myself to mimicking her patterns of speech, which got me nicknamed '*La française*' in the schoolyard. As soon as my homework was done I'd wiggle through a hole in the honeysuckle hedge and scratch at her back door.

Geneviève was in her late twenties, childless and homesick; she'd followed her husband as he was transferred from country to country, lugging a battery of Le Creuset pots and pans and a closetful of pastel dresses in swirly psychedelic or whimsical floral patterns which she'd happily model for me, and sometimes let me try on. I'd clack around in her pumps and twirl in front of the mirror. Sometimes we'd both dress up and stage make-believe photo shoots inspired by *Vogue*. Those were grand occasions since Geneviève would also let me pick from the array of cosmetics on her dressing table and carefully do my face. We'd model the looks in our very favourite makeup ads, the ones for Dior: pale, moody, smoky-eyed beauties with thin scarlet lips.

But there was one particular item on the dressing table I steered well clear of: a blue, black and silver-striped canister that said 'Yves Saint Laurent Rive Gauche'. What if my lungs seized up? I hadn't suffered an asthma attack since the age of six, but I'd witnessed my dad's discomfort if we walked within ten feet of the perfume counters in the local shopping mall, so I wasn't taking any chances. Geneviève gave a Gallic shrug when I finally, cringingly, explained about the allergies.

'You North Americans really indulge your little *bobos*, don't you? Here, look at this …'

She set her smouldering Camel in an ashtray, pulled out the scrapbook where she kept magazine cuttings and pointed to the picture of a slender young man with huge square glasses flanked by a lanky blonde in a safari jacket and a cat-eyed waif with a gypsy scarf on her head. The trio exuded a loose-limbed pop-star glamour. This was, Geneviève explained, Yves Saint Laurent, the greatest couturier in France, with his muses Betty Catroux and Loulou de la Falaise. And Rive Gauche was the perfume he'd named after his new boutique in Saint-Germain-des-Prés, the first place she'd head for when she went back to Paris.

From the ads I'd spotted in my mother's *Good Homemaker* magazine, I knew perfumes ought to have fancy glass bottles and evocative names like Je Reviens or Chantilly. There was nothing poetic about that metal canister. And Rive Gauche, what kind of a name was that? So Geneviève showed me the ad for Rive Gauche: a redhead in a black vinyl trench coat strolling by a café terrace with a knowing smile. *Rive Gauche, plus qu'un comportement*, it said; *Rive Gauche, un parfum insolite, insolent.* 'More than an attitude. An unusual, insolent perfume.'

My friend explained about the Paris Left Bank, the jazz clubs and bohemian cafés on the boulevard Saint-Germain she used to walk by as a teenager to catch a glimpse of *les philosophes*, Jean-Paul Sartre and Simone de Beauvoir. I'd read about Socrates in my children's encyclopaedia but hadn't fathomed that there were living philosophers or that they could even remotely be thought of as cool. At my age, cool was still a difficult concept to grasp.

'Like pop stars, you mean?'

Geneviève nodded. Yves Saint Laurent, she went on, expressed the spirit of the Left Bank: youthful, rebellious and free. I couldn't quite figure out how the soapy, rosy-green scent that clung to Geneviève's clothes could reflect notions like youth, rebellion, freedom or insolence, and I had no idea of what constituted an 'unusual' fragrance. But I did half-guess from her wistful gaze that Geneviève, trapped in a Montreal suburb where there were no cafés haunted by chic bohemian philosophers or couturiers – there weren't even any sidewalks! – longed for that lost world. A world captured in that blue, black and silver canister …

One afternoon, our orange school bus dropped us off early so that we could prepare for the year-end recital: I sang in the choir, humiliatingly tucked away in the last row with the boys because I'd suddenly grown taller than all the girls. Geneviève had promised that, for the occasion, she would do my hair up in her own signature style, a complex hive of curls fastened with bobby pins. I scrunched my eyes and held my breath as Geneviève stiffened her capillary edifice with spurts of Elnett hairspray.

'There you go … Have a look!' she said, waving a *Vogue* around to clear the fumes.

I opened my eyes to a chubby-cheeked version of Geneviève. The chignon was practically as high as my head: I'd tower over the boys too.

'And now …' Geneviève reached for her Rive Gauche. 'As a special treat, I'll let you wear my perfume … In France, an elegant young lady never goes out without fragrance.'

That spritz of Rive Gauche didn't kill me; in fact, it made me feel better than I'd ever felt, so grown-up, so *important* – an eleven-year-old girl with the newest perfume from Paris! That's when I resolved that when I grew up, I'd be like Geneviève, and never leave the house without a drop of perfume. And it would be French perfume, even if I had to swim across the Atlantic to get it.

4

What is it about the French and perfume? Draw up a list of the greatest perfumes in history. Shalimar, Mitsouko, N°5, Arpège, Femme, L'Air du Temps, Diorissimo? French. Study the top ten sellers in any given country. The labels may be American, Italian or Japanese, but the perfumers who composed them? At least half are French and most of the others are French-trained.

When Bourjois, the cosmetics company that owned Chanel perfumes, decided to put out a fragrance called Evening in Paris in 1928, they knew full well that they were launching the ultimate aspirational product. For millions of women, that midnight-blue bottle would hold the prestige and romance of the French capital within its flanks – it was the closest most would ever come to the Eiffel Tower. Judging from the number of Evening in Paris bottles that keep popping up on auction websites, they were right. For generations, 'French perfume' was the most desirable gift, short of mink and diamonds, and a lot more affordable.

But why is it that those two words, 'French' and 'perfume', have been said in the same breath for centuries? In other words:

why is perfume French? If you ask most people in the industry, they'll answer, 'Well, because of Grasse, I guess,' Grasse being the town in the South of France where perfumery developed as an offshoot of the leather-tanning industry. Tanning products were rank, so fine leathers were steeped in aromatic essences to counteract the stench, and Grasse enjoyed a particularly favourable microclimate for growing them. Though most of the land has now been sold to real-estate developers, it is still very much a perfumery centre, with several labs and a few prominent perfumers based in the area. But 'Grasse' doesn't answer the question. There were other places in the world, like Italy and Spain, where a cornucopia of aromatic plants could be grown; where botanists, alchemists and apothecaries studied them, refined extraction processes, experimented with blends. There must be another reason why it was in France that perfume went from a smell-good recipe to liquid poetry; why it was here and nowhere else that modern perfumery was born, thrived and gained international prestige.

So why indeed? If anyone can answer the question I've been asking myself since I was eleven, it is the historian Elisabeth de Feydeau. We've just been enjoying an al fresco lunch in a garden gone wild with roses, lush with vegetal smells rising in the afternoon heat. A tall, chic blonde with a sweet, sexy-raspy voice, Elisabeth was formerly the head of cultural affairs at Chanel; she teaches at ISIPCA, the French school of perfumery, as well as acting as a consultant for several major houses; she wrote a book about Marie-Antoinette's perfumer and a history of fragrance. So as we nibble on petal-coloured cupcakes, Elisabeth graciously shifts into teaching gear. I have indeed come to the right place for an answer, she tells me; the sacred union between France and fragrance was sealed right here where we're sitting, in Versailles.

If Catherine de' Medici hadn't come to France in 1553 to marry the future King Henri II, perfume might well have been Italian. Not only did Italy enjoy a climate allowing the cultivation of the plants used in perfumery, but with Venice lording it over the sea routes, all the precious aromatic materials of the Orient flowed into the peninsula. And in the dazzlingly refined Italian courts of the Renaissance where the young Duchess Catherine was raised, perfume-making, intimately linked to alchemy, was a princely pastime practised by the likes of Cosimo di Medici, Catarina Sforza and Gabriella d'Este. Italian alchemists had started to divulge their methods of distillation and many of their perfumery treatises had already been translated. When Catherine de' Medici arrived with her perfumer Renato Bianco in tow, she brought along the Italian tradition in all its refinement.

The French perfume industry was centred in Montpellier, where research on aromatic substances and distillation was carried out at the faculty of medicine (one of the oldest in the world, founded in 1220), and in Grasse, where skins imported from Spain, Italy and the Levant were treated. Up to then, perfumery had remained a subsidiary activity for apothecaries and tanners. Spurred on by the Italian fashion for scented clothes, leather items, pomanders and sachets, it developed into a luxury trade.

But the true turning point came from the scent-crazed king who had determined to transform his court into the crucible of elegance: it was under the reign of Louis XIV (1638–1715) that the French luxury industries acquired the excellence and prestige they still boast today. And that spectacularly successful marketing operation was a very deliberate political endeavour … During his mother Anne of Austria's regency the young king had lived through the War of the Fronde, an uprising of the nobility that had threatened the very existence of the monarchy. To keep his noblemen under control, Louis XIV decided to move his court away from Paris to the newly built Versailles. There, he transformed his most trivial activities – getting out of bed, being groomed and dressed and even using the 'pierced chair' (the 17th century version of the toilet, known in England as 'The French

Courtesy') – into a series of ritual displays which courtiers had to attend to curry favour with the monarch. By compelling them to follow the fashions he launched and to participate in the court's lavish spectacles, the Sun King made sure they had no time or money to plot against the Crown. To imitate him, courtiers turned their *toilette* (named after the *toile*, the piece of cloth onto which cosmetics were spread out) into a social occasion. Celebrity endorsement was invented in Versailles. If, say, the king's current favourite used some new pomade or scented powder in front of her coterie, she might create a stampede for it. This bit of information would be carried by a brand-new medium, the fashion gazette.

This was the other reason behind Louis XIV's marketing operation. Fashion would radiate from his royal person to his court, from the court to Paris, and from Paris to the entire world, which would draw much-needed currency into the kingdom. His minister of finance Jean-Baptiste Colbert therefore set out to build up the French luxury industry, sending spies abroad to steal trade secrets and poach skilled artisans, and finding out which professional corporations were likely to become economic forces for France. Colbert made a deal with the guilds: they would be granted privileges and exempted from certain taxes provided they came up with the best and the most beautiful products. 'The château of Versailles became, quite literally, a showroom designed to display the know-how of French craftsmen to foreign monarchs and dignitaries. There was an order book when you came out!' chuckles Elisabeth de Feydeau. Paris, with its fancy shops staffed by well-turned-out young women, glittering cafés and public gardens, became an extension of that Versailles showroom, spurring on further demand for made-in-France luxury items, which in turn spurred on creativity, as perfumers refined their art to meet the exacting tastes of their snobbish clientele.

The perfume and cosmetics industry was one of those encouraged by Colbert. He boosted its profitability by setting up the Company of the Eastern Indies to ensure the supply of exotic ingredients while plantations were developed in Grasse. Again, Louis XIV provided the best possible celebrity endorsement: his orange blossom water, which came from the trees in the Versailles orangery, was exported all over the world. It was the only perfume the Sun King tolerated in his later years: in the last decade of his reign, perfumes gave him migraines and fainting spells which may have been psychosomatic, or due to allergies. 'No man ever loved scents as much, or feared them as much after having abused of them', wrote the memoirist Saint-Simon.

In fact, the royal scent-phobia just about killed the industry. A Sicilian traveller noted that 'foreigners enjoy in Paris all the pleasures which can flatter the senses, except smell. As the king does not like scents, everyone feels obliged to hate them; ladies affect to faint at the sight of a flower.' And indeed, though Versailles and Paris reverted to their old fragrant ways under the gallant reign of Louis XV, one of the most groundbreaking products in the history of the industry came not from Grasse, Montpellier or Paris, but from Italy via Cologne, Germany, where Johann Maria Farina launched his Aqua Mirabilis. The 'Miraculous Water', a light, bracing blend of citrus, aromatic herbs and floral notes in an alcoholic solution, is known to this day as 'eau de Cologne'. This new style of perfumery, a departure from the heavy, animal notes used by Louis XV's libertine court, was well in tune with the tentative progress of personal hygiene practices and the tremendously fashionable back-to-nature philosophy of Jean-Jacques Rousseau. Noses and sentiments were becoming more delicate …

As Rousseauism swept through Europe, clothing, interiors, gardens and fragrances took on the tender floral tones that most flattered the fresh complexion of the kingdom's premier fashionista, Queen Marie-Antoinette. It was under her reign that Paris consolidated its status as the trend-setting capital of the world. By the time her husband Louis XVI was crowned in 1774, the rigid etiquette of Versailles set up by the Sun King had lapsed and, as a consequence, the status of the aristocracy was rather less exalted than it had been. To stay ahead of the rising bourgeois class, the ladies of the court, led by Marie-Antoinette, resorted to the only thing that could keep them one step ahead of the commoners, however wealthy they were: fashion. In fact, this is how fashion as we know it – the latest trend adopted by a happy few for a season before trickling down to the middle classes – came into existence. Marie-Antoinette plotted the newest styles with her 'Minister of Fashion', Rose Bertin. She and her entourage would retain exclusivity for a set period, after which the item could be sold in Bertin's Paris shop, Le Grand Mogol, and every woman who could afford it could dress herself '*à la reine*'. This, in itself, was a revolutionary move. As Elisabeth de Feydeau explains it, etiquette demanded that the Queen use Versailles suppliers. But Marie-Antoinette wanted her suppliers to go on living in Paris to sniff out the zeitgeist. Louis XIV had established the prestige of the French luxury trades, 'but that little twist you can only find in Paris, that *je ne sais quoi* that makes Parisian women incomparable – and this comes out very clearly in the writings of foreign visitors of the period – emerges in the 18th century. There is a word that sums this up: elegance.'

With her gracious manners and lively laugh, Elisabeth herself is the very type of *Parisienne* foreign visitors have admired since Marie-Antoinette's day. Even the dishevelled garden where we converse has just a touch of the *négligence étudiée* that distinguishes chic Parisian women from their fiercely put-together New Yorker or Milanese counterparts. I can easily envision Madame de Feydeau in one of the supple white muslin gowns the queen made fashionable in the 1770s, presiding over an assembly of philosophers and artists with whom she could match wits. Because this is what Paris gave to the world, through the learned, elegant women who led its social life and were deemed worthy intellectual sparring partners by the brightest men of the age: wit rather than pedantry; a way of thinking of life's pleasures with the finely honed weapons of philosophy. You've always wondered why the French appreciate smart women? Why women don't drop off the radar after the age of forty? There you have it. Where the *art de vivre* is a serious pursuit, the women who have turned it into an art form are valued indeed. And this, we owe to the 18th century.

Thus, it isn't surprising that Elisabeth turned her attention to that very period with *A Scented Palace: The Secret History of Marie-Antoinette's Perfumer*. Jean-Louis Fargeon rose to the top of his profession just as perfumery was severing its ties with tannery and glove-making to become a fully fledged trade – another turning point in the history of the industry. If he left his native Montpellier, it was because he knew that he could only succeed if he breathed in the incomparable *esprit de Paris*. And if he nabbed the world's most prestigious customer, it was because he was incomparably skilled at composing the delicate floral blends that reflected the era's craving for a simpler, more natural life … Fargeon's status as the queen's perfumer almost cost him his head during the French Revolution: he escaped the guillotine by a hair's breadth simply because, as he was rotting in jail, Robespierre was executed, thus ending the Terror that cost the lives of over forty thousand people, a full third of whom were craftsmen. And the Revolution did, in fact, almost kill off the French perfume industry, not only because it dissolved the guild of perfumers, which reeked of aristocratic privilege, but because most of its clients were either beheaded or exiled. By the time a new court was formed around the Emperor Napoleon, tastes had changed. Though the Empress Josephine was inordinately fond of musk, he preferred her natural aroma, famously asking her not to wash for several days because he was returning from a military campaign. And he loathed perfume apart from eau de Cologne, which he carried everywhere with him, literally showered with and even consumed by dunking lumps of sugar in it. After his fall, the French luxury trades declined while the English industry gained ascendancy.

'Then, in 1830, French perfumers came up with the phrase "Invent or perish",' reveals Elisabeth. 'They realized that the world had changed, that the perfumery of the old regime was no longer possible, and from 1840 onwards, French perfumery started rebuilding itself.'

But the period was dominated by bourgeois, puritanical values, and vehemently rejected the wastefulness and deceit of cosmetics and fragrances. 'Perfumes are no longer fashionable,' sniffed a French Emily Post in 1838. 'They were unhealthy and unsuitable for women for they attracted attention.' At most, a virtuous woman could smell of flowers, and not just any flowers: heliotrope, lilac, carnations or the whitest possible roses. Master perfumers like the Guerlains – the house was founded in 1828 – undoubtedly made lovely blends for their wealthy customers, but the bulk of scents were mainly used as personal hygiene products in an era when most homes didn't have running water. Fragrance was no longer a luxury: it was just a way of smelling nice, of removing bad smells.

Paris, however, remained the capital of fashion: it was here that the Englishman Charles Frederick Worth established himself as the world's first star couturier, under the reign of Napoleon III. But French perfumery would have lagged behind without the handful of visionaries who propelled it into the 20th century.

One of these men was Jacques Guerlain, who authored a series of shimmering, impressionistic masterpieces still produced and adored to this day: Après l'Ondée, L'Heure Bleue, Mitsouko and Shalimar. But an upstart was hot on his heels, a self-taught perfumer untrammelled by tradition and therefore willing to use the new, more brutal synthetic materials to conjure vibrant Fauvist compositions in tune with the scandalous Ballets Russes. It was the Corsican François Coty, known as the Napoleon of perfumery, who democratized fine fragrances by launching cheaper ancillary lines, whereas Guerlain perfumes were still sold to a very select clientele. Soon, he was selling 16 million boxes of talcum scented with L'Origan in France alone. Guerlain had reinvented the heritage of traditional perfumery: Coty turned it into a worldwide industry and took it into the 20th century.

But it was haute couture that truly transmogrified Paris perfumes into the stuff dreams were made of for millions of women around the planet – the stuff that would make a young girl in the suburbs of Montreal decide she would be Parisian one day because of a spritz of Rive Gauche. 'As long as you're in a one-to-one relationship with your customer, like Jean-Louis Fargeon was with Marie-Antoinette, or Jacques Guerlain with the clients who came to his salon, you can explain your perfume, make it loved,' Elisabeth tells me. 'But when fragrance becomes an industry, how do you sell it as a luxury item? Perfume needs to be supported by image. And who but world-famous Paris couturiers could provide a better one?'

The first couturier to launch a perfume house was Paul Poiret, in 1911. The Pasha of Paris, as he was dubbed, was practically the first to turn fragrance into a concept, with a dedicated bottle and box for each product. But he made the marketing mistake of not giving it his own name. The first to do that, in 1919, was Maurice Babani, who specialized in Oriental-style fashions and whose perfumes bore such exotic names as Ambre de Delhi, Afghani or Saigon. Gabrielle Chanel followed suit in 1921, but her N°5 only went into wider production in 1925, when she partnered up with the Wertheimers who owned Bourjois cosmetics. By that time, other couturiers like Jeanne Lanvin and Jean Patou had realized perfume was a juicy source of profit and got in on the game. Thus it was that in the first decades of the 20th century, French couture and perfumery began an association from which the prestige of the most portable of luxury goods – a few drops on a wrist or behind the ear – would durably benefit.

But it wasn't all a canny exploitation of the Parisian myth. French perfumery truly *was* the best and the most innovative; French perfumers *did* create most of the templates from which modern perfumery arose. Why? Perhaps because, ever since

Colbert's day, Paris had been a laboratory of taste, not only in fashion or perfume, but also in cuisine and decoration. A discerning clientele of early adopters, eager to discover new styles and new fashions, spurred the purveyors of luxury goods into devising ever more refined, more surprising, more delightful products, thus bringing their craft to an incomparable degree of perfection. Function, as it were, created the organ. In this case: the best noses in the world.

5

However, my very first perfume was not French, but American. Its frosted apple-shaped bottle has managed to survive the decades without getting carried off in my mother's garage sales. It followed my parents when they sold the house where I grew up. It still taunts me whenever I go home to visit ...

Geneviève certainly meant well when she picked Max Factor's Green Apple for me as a parting gift before she followed her husband to Saskatchewan. She must have been told it was suitable for young girls. I hated it. Unlike Rive Gauche, which had given me a glimpse of the woman I aspired to become, Green Apple didn't tell a story. Or if it did, it was a story I didn't want to hear in my first year in high school; a story that contradicted my training bra and the white elastic band that had already cut twice into my budding hips, holding up Kotex Soft Impression (*Be a question. Be an answer. Be a beautiful story. But be sure.*) What it said was: 'Don't grow up.'

Wear Green Apple: He'll bite, insinuated the Max Factor ad. Never mind that apple candy has never registered as a powerful male attractant and that I'd never been kissed by a 'he', much less bitten. I'd attended Sister Aline's catechism class and I *got* the

Garden of Eden reference behind that slogan, thank you. But to me, that bottle wasn't the fruit of the Tree of Knowledge. It was Snow White's poison apple. And an unripe one at that, just like me. The only apple-shaped things I wanted to own were starting to fill my A-cups.

When the good people at Max Factor put out their apple-shaped bottle, conjuring memories of Original Sin and of the Brother Grimms' oedipal coming-of-age tale, they tapped into a symbolic association which Dior would later fully exploit with another potion in an apple. But that one would be called Poison: the myth of the femme fatale poured into amethyst glass and wrapped in a moiré emerald box – purple and green themselves hinting at venom and witchcraft …

When it was launched in 1985, I'd long grown out of my Green Apple trauma and into the twin addictions of ink (usually purple) and perfume. I hoped to become a writer, and a fragrant one at that. I'd just finished writing my Masters dissertation on 18th century literature and I was struggling through my first academic publication in France, an essay on Flaubert's *Madame Bovary*, when I stumbled on an article about Poison, 'a spicy perfume blending Malaysian pepper, ambergris, orange blossom honey and wild berries'. This was particularly serendipitous: the whole point of my essay was that the adulterous Emma Bovary's aspiration to become a romantic heroine despite the mediocrity of her provincial surroundings was the result of her mind being poisoned by cheap romantic novels. A farmer's daughter married off to a small-town doctor, she grew so frustrated at not being able to live out her girlhood fantasies that 'she wanted to die. And she wanted to live in Paris.' Emma never got closer to Paris than Rouen but she managed to run up such a debt with the local frippery pedlar that her house and belongings had to be auctioned off, whereupon she poisoned herself with arsenic. In my essay, I contended that the ink she had absorbed when reading romances had poisoned her every bit as much as the rat-killer.

'What I blame [women] for especially is their need for poetization,' Flaubert wrote in 1852 to his lover Louise Colet. 'Their common disease is to ask oranges from apple trees … They mistake their ass for their heart and think the moon was made to light their boudoir.' As neat a forecast as ever was of what would drive perfume advertising one century hence. Weren't we all modern embodiments of Emma Bovary, I thought, seeking magic potions to transform our lives? To die or live in Paris … And now, a legendary couture house where a latter-day Bovary could well burn through her husband's earnings was selling an intoxicating – and perhaps toxic – dream of romance and glamour under the slogan 'Poison is my Potion.'

When the CEO of Christian Dior Parfums, Maurice Roger, first decided, in 1982, to launch Poison, the industry had been undergoing deep mutations. Perfume houses were being bought up by multinational companies, and their new products had to have global appeal. With the blockbusting, take-no-prisoners Giorgio Beverly Hills, America was starting to bite into a market that had hitherto been dominated by France. To make itself heard around the world in the brash, loud, money-driven 80s, the house of Dior had to create a shock. This, Maurice Roger knew, meant renouncing the variations on the name 'Dior' which had been used to christen all their feminine fragrances from the 1947 Miss Dior to the 1979 Dioressence; it meant foregoing Parisian 'good taste' and striking with an unprecedentedly powerful scent and concept. It meant re-injecting magic, mystery and a touch of evil into something that was no longer for 'special moments' (think of that fragrance your grandmother kept on her dressing table and only dabbed on when she pulled out the mink and pearls), no longer a scent to which women were wedded for life, but a consumer product. The magic, this time, would be entirely marketing-driven, though the fragrance itself, selected out of eight hundred initial proposals, required over seven hundred modifications before reaching its final form.

The promotional campaign kicked off with a 'Poison Ball' for 800 guests at the castle of Vaux-le-Vicomte hosted by the French film star Isabelle Adjani. And, if the bottle was reminiscent of the story of Snow White, the advertising film conjured another fairy tale, The Beauty and the Beast. Its director, Claude Chabrol, never hinted that the would-be femme fatale might be the one to succumb, Bovary-style, to the potion it touted. But I'm sure Chabrol, who would go on to direct an adaptation of *Madame Bovary*, had a thought for one of his film heroines, the eponymous and fragrantly named Violette Nozière, a young prostitute found guilty of poisoning her parents in 1934 (both characters played by the exquisitely venomous Isabelle Huppert).

'Who would offer someone a gift of poison?' the *New York Times* fashion editor wailed. And Dior's blockbuster did, in fact, wreak havoc in its wake. Innocent bystanders complained of headaches and dizziness; diners were put off their food by its near-toxic intensity. It was banned in some American restaurants, where signs read 'No Smoking. No Poison.'

Perfume is an ambiguous object, forever hovering on the edge of stink: what smells suave to me can be noxious to you, and the padded-shouldered juices of the 80s gave the first ammunition to the anti-perfume movement. Google 'perfume + poison' today and you're just as likely to find entries about Dior as exposés on the purported toxicity of fragrance. Perfume aversion is no longer fuelled by the strength of the smell of perfume but by fear of the invisible chemicals that might be infiltrating our bodies through our skin and lungs. Activist groups have been flooding the internet with ominous warnings about 'unlisted ingredients' which sound as if they belong in some mad scientist's lethal cocktail and would be enough to put off anyone but the staunchest fragrance aficionado. And since perfume, unlike anthrax, radiation, or the toxins in building materials, food and water, is both perceptible and dispensable, it's crystallized a rampant paranoia. If you start obsessing that the woman in the next cubicle is poisoning you with her Obsession, at least you can do something about it: file a lawsuit.

But the fear of smells itself, brilliantly explored by the French historian Alain Corbin in his 1982 classic *The Foul and the Fragrant*, goes back millennia. Until the link between germs and disease was established in the late 19th century, people were convinced smells could kill. As far back as the 5th century AD, physicians were accusing foul odours of causing plagues. Since nobody knew how illnesses spread, it seemed logical to suspect the airborne stench rising from graveyards, charnels, open cesspools or stinky swamps. And no less logical to suppose that the stench, and hence the disease it carried, could be repelled by another, stronger smell. Aromatic materials were not only considered salubrious because of their medicinal properties: they were also believed to be the very opposite of the corruptible animal and vegetal matters whose putrescence was thought to cause disease. In fact, they were just about as pure as earthly matter could be, since aromatic resins burned without residue, and essential oils were the volatile spirit that remained after the distillation of plants. Those very same aromatic materials were used to prevent corpses from putrefying. Could they not prevent the corruption of the living body by disease as well?

So 'plague doctors' stuffed them in beak-shaped masks as they made their rounds during epidemics of the Black Death; aromatic materials were burned in houses, churches, streets and hospitals. And if the wealthy lavishly scented their clothes and held pomanders to their noses, it wasn't only to cover up the effluvia of the great unwashed or the stench of cities and palaces: scent was their invisible armour against the Reaper. In fact, if everyone from kings to paupers *did* give off quite a pungent aroma from the late 15th century, when public baths were shut down for moral reasons (they often harboured prostitutes), until the late 18th century, it was precisely *because* foul miasmas were believed to carry disease. Physicians were convinced that water distended the fibres of the body, leaving it wide open to airborne contamination. Therefore, bathing was undertaken only under strict medical supervision, and people who'd just bathed were advised to stay under wraps for hours, if not days, until their bodies had sufficiently recovered to withstand exposure. Such cleanliness as there was came from washing face, teeth and hands with various cosmetic preparations; bodily grime was meant to be absorbed by white linen shirts, which only the upper classes could afford to change daily.

But the very notion that perfume had curative or protective properties implied that it could have the reverse effect, at least potentially. In Ancient Greece, the word *pharmakon*, from which 'pharmacy' is derived, designated both the poison and the remedy. If aromatic essences could act on the body by penetrating through the nose or skin, they might also kill. Odours are invisible: unlike poisonous foods which you can choose not to eat if you suspect something is amiss, you can't defend yourself against them – you can't hold your breath forever. And strong, apparently pleasant fragrances could very well conceal subtle poisons.

In the midst of the religious war that tore France apart in the late 16th century, the Catholic Queen Catherine de' Medici was thus suspected of having assassinated the Protestant Jeanne d'Albret, Queen of Navarre, with a pair of poisoned gloves made by her perfumer Renato the Florentine. Though unfounded (Queen Jeanne died of tuberculosis), the accusation gives an indication of the ambivalent properties attributed to perfume. Even natural fragrances were suspicious. In 1632, in a clear-cut case of collective hysteria, the nuns of a convent in Loudun accused the priest Urbain Grandier of having cast a spell on them: the first to be 'possessed' by the charismatic Grandier claimed she had been bewitched by the smell of a bouquet of musk roses. Grandier was burned at the stake for witchcraft.

By the middle of the 18th century, some of the very aromatic materials that had been considered both exquisite and prophylactic were being condemned. Substances of animal origin such as musk, civet and ambergris were lumped in with putrid matters through medical anathema. The smell of musk, for instance, was compared to that of manure or fermented human excrement. Its very strength unsettled 'our more delicate nerves', wrote Diderot and d'Alembert in their *Encyclopaedia* in 1765. As the upper classes renewed their acquaintance with water and came to enjoy light, vegetal scents such as the eau de Cologne, wearing heady animalic concoctions became the sign not only of doubtful hygiene but of depraved tastes, fit only for skanky old libertines and their whores.

Perfume would remain a remedy until 1810, when Napoleon decreed separate statuses for perfumers and pharmacists: since the latter were compelled to disclose the composition of their preparations, the former chose to relinquish any medicinal claims to preserve their trade secrets. But perfumers never completely shook off the ambivalence of the original *pharmakon* … No wonder a perfume-phobic pharmacologist's daughter ended up sticking her nose in it.

If I wanted further proof that perfume *is* indeed toxic, I'd filch my dad's old lab coat to analyse the 1999 Hypnotic Poison, whose red bottle contains the antidote to the green one I was given over twenty years ago: Snow White's poisoned apple, ripe and fit at last for a grown woman.

Was the perfumer Annick Menardo aware of what she was doing when she stuck an almond note into its jasmine sambac, musk and vanilla accords? As any reader of classic English murder mysteries knows, you can tell whether a victim has been poisoned with cyanide from the lingering smell of bitter almonds. In fact, like apple seeds or peach and cherry pits, the bitter almond contains a highly poisonous substance, amygdalin, which in turn contains sugar and benzaldehyde, a common aromatic ingredient that smells like amaretto liqueur, but can also yield cyanide. And thus, Hypnotic literally reeks of poison disguised as a delicacy, its toxicity betrayed by the slight bitterness of caraway seeds rising from a powdery cloud …

I'd long wanted to ask Menardo about that almond note, and lay my love at her feet for what must surely be one of the weirdest scents to come out in a mainstream brand, Bulgari Black. When I wear Black, I feel that I've either a) dropped my liquorice macaroon in my cup of lapsang souchong, b) powdered my butt with fancy talcum and slipped on my rubber bondage skirt, c) crossed a tough neighbourhood where someone's been burning tyres in a cab that's got one of those little vanilla-scented trees dangling from the rear-view mirror, or d) been guzzling the world's peatiest single malt whisky and gargled with Shalimar to hide the fact.

But getting in touch with the brilliantly gifted Menardo was a daunting task. Perfumers working for big companies are much less accessible than independents like Bertrand Duchaufour. You can't just cold-call them – least of all Menardo, who is famously reticent and uncompromising. Every time I brought her name up, insiders would wish me luck. And I did get lucky: while visiting the offices of Firmenich, where she works, I bumped into her in a corridor, a tiny bristling dark-haired sprite in a hot-pink top. It seemed awkward to spring my question about almonds – how pleasant can it be to be waylaid by some strange woman on your way to a meeting or to the loo? So after gushing about Bulgari Black (she gruffly responded, 'But that's old stuff!'), I let her go, contacted her again through email, and we made a phone appointment so that I could ask her about Hypnotic Poison.

So: *did* she work in the almond note on purpose? She chuckles. 'I didn't psychoanalyse myself. It's possible.'

She explains she worked on the idea of toxicity by playing on the *strength* of the smell: 'I knew I had to do something that was meaner than Poison.'

I point out to Annick that the apple seems to be a leitmotif in her career: another one of her best-sellers, the anise and violet Lolita Lempicka, which came out the same year as Hypnotic Poison, is also packaged in an apple-shaped bottle. She answers that apple itself is one of her fetish notes – a fond memory of a shampoo she used as a teenager, Prairial – so she sticks it in wherever she can. By this time we're chatting away quite happily, and I confide the story of Geneviève's poisoned parting gift.

'Green Apple? Hey, I wore that too when I was a kid!'

My phone almost drops from my hand. The formidable little witch who whipped up my antidote to Green Apple actually wore it herself. There are no coincidences with perfume: it's all black magic.

6

Despite my admiration for the cavity-inducing Hypnotic Poison, olfactory pastries were never something I could get particularly worked up about. I may want to offer myself up at dessert when the mood strikes; I don't want to smell of it. Which is why Bertrand Duchaufour's take on vanilla delights me particularly: it reminds me of something I'd much rather wrap my lips around after dinner than a spoonful of vanilla ice cream … a good cigar.

Surprised? Yes, the vanilla pod, as opposed to the synthetic vanillin more frequently used in fragrances and desserts, *does* have a tobacco facet, and that's what Bertrand has chosen to underline, down to the slight vegetal mustiness of cured tobacco leaves. This is one of the things I find most intriguing about his style: the way he slips in weird notes that mess up the *prettiness* of a scent. For instance, the quasi-surrealistic way he grows a Cuban cigar out of a vanilla pod, as though that had been the pod's subconscious desire all along.

Bertrand's been making good on his promise: this is my third lesson in three weeks. Today I've asked him to explain the structure of one of his compositions so that I can re-enact the demonstration in my London course: I'm making good on my promise too, since I told him I'd include his work in the syllabus. I'd be a fool not to. After all, the man is one of the most distinctive perfumers in the business, one of the few who has enough artistic liberty to develop a consistent oeuvre and to impose his vision on the projects he takes on.

It took balls to stake that claim. He had to break out of a system that was set up in the late 30s and still dominates the industry, similar in its set-up to the system of the classic Hollywood era, with directors on the payroll of studios and forced to work within their constraints. Like the overwhelming majority of his colleagues, Bertrand was thus employed by Symrise, one of the big labs that produce aromatic materials and compositions.

He'd wanted to be a perfumer since the age of sixteen, when a girlfriend introduced him to Chanel N°19, but he'd been advised to skip the Versailles perfumery school and get an internship in Grasse. As he had no contacts there, he gave up, convinced he'd never become a perfumer. After studying biochemistry in Marseilles, he took a gap-year to tramp around South America before coming back to France to study for a degree in genetics. It was then he learned that a friend of his had won the coveted internship in Grasse: at last, he'd found the contact he needed. He got one too, stayed on, and ended up in the Paris branch of the company, where he was mentored by Jean-Louis Sieuzac, who authored such best-sellers as Opium for Yves Saint Laurent, Dune and Fahrenheit for Christian Dior or Jungle Elephant for Kenzo. In time, he made his way up to senior perfumer.

Being a perfumer on the payroll of a big lab can be a thankless job. Fine fragrances are the most prestigious gigs but, more often than not, you're put to work on functional fragrances, the stuff that goes into detergents, cosmetics or hygiene products, which is where the big money is because of the volumes involved. It's technically challenging and it can be aesthetically gratifying – just smell Ajax Spring Flowers and tell me if it's not as good as some of the juices that are sold in department stores – but it's certainly not glamorous.

Fine fragrance perfumers don't necessarily get much more wiggle room because of the way the system is set up. It goes like this: the client, usually a designer brand, wants to put out a new fragrance. The brand's marketing team comes up with the name, the bottle design and the concept for an ad campaign before anyone remembers that something actually needs to go into that bottle. A brief specifying the style, target market and cost of the product is knocked together and handed out to several competing labs. The staff perfumers who are interested in the project must come up with proposals, usually within a matter of a few weeks. The budget rarely goes over sixty to eighty euros per kilo of oil (the blend of aromatic materials to which alcohol will later be added). By way of comparison, the price per kilo for niche perfumes can shoot up to four hundred or even six hundred euros, but what actually ends up in a department-store bottle is worth less than ten euros. The labs present their proposals, which they develop on spec. The client selects one. The perfumer goes back to the lab to tweak it; sometimes several team up to accelerate the process. If it is a big project, the product is tested on consumer panels, after which it is tweaked some more, often until every original molecule has been blasted out of its body. If sales don't do well, it may be tweaked again.

The system is hardly conducive to creativity. Perfumers learn to do compositions that test well in order to win the brief: after all, they've been hired to make money for their employers. What tests well is what consumers are familiar with. What consumers are familiar with are either best-selling fragrances or everyday products like shower gels, fabric softeners and shampoos. Recipes that sell well get around from one perfumer to another and from one company to another; they are recycled endlessly, so that you find the same accords in every fragrance. The same twenty to thirty raw materials are used over and over, out of the thousands that exist. On top of that, the systematic use of gas chromatography, a method that allows companies to analyse the competition's products, has led to a practice called the 'remix': take current best-sellers, cut and paste, and you'll have the next designer-brand juice. Another practice is called the 'twist': take a best-seller, change a couple of things in it and presto! Pour it into a bottle. If you ever wondered why everything smells the same in department stores, now you know.

Like so many of his colleagues, Bertrand became a perfumer because he was fascinated by the classics. The market forced him to go in another direction, though from the late 90s onwards he was lucky enough to work with people who did value originality and afforded him an opportunity to develop his distinctive style. But he was too much of a maverick to fit into the 'studio system' for ever, which is why he and his employers eventually decided to part ways by common accord.

'I was asked to do things that didn't interest me and I had a lot of trouble coming up with a decent product both for the company and for the brand.'

'So I guess you weren't offered a lot of stuff to do …'

'I wasn't offered anything any more.'

'Because you were difficult?'

'Because I told them to bugger off.'

'In other words, you were a pain in the ass.'

'I was a pain in the ass, that you can be sure of!' he chuckles.

I'm quite sure Bertrand can be a pain in the ass, and I can readily envision his temper flaring up if he's prodded too hard. That's why I'm a little wary of asking him if he was serious when he said my story would make a great perfume. What if he brushes me off? I'm hoping he'll bring the subject up himself, but right now he's busy playing show-and-tell with vanilla absolute, talking me through its facets in a gleeful, earnest voice, as though he were rediscovering it all over again. Vanilla has animal, leathery and smoky facets, he enthuses; it also has woody, ambery, spicy and balsamic facets, and even unpleasant medicinal notes. That's what he wanted to get at: to draw out every aspect of the vanilla pod.

In the scent, the vanilla acts as the core of a star-shaped structure. Its different facets are picked up and amplified by the other materials, to form a second, phantom vanilla; an olfactory *illusion* sheathing the real thing; a space in which all the notes resonate.

As Bertrand speaks, I scribble a diagram with vanilla as the 'sun' and the other materials as 'planets': rum, orange, davana (fruity, boozy), immortelle (walnut, curry, maple syrup, burnt sugar), tonka bean (hay, tobacco, almond, honey) and narcissus (hay, horse, green/wet, floral). Pretty soon Bertrand is scribbling in my notebook too, writing down the effects conjured when the different materials meet. For instance, rum and immortelle emphasize the woody/ambery facets: because rum is aged in oak casks, it already has a vanilla flavour imparted by the oak (vanillin can be synthesized from by-products of the wood industry). It all ties in: the sheer *logic* of it is limpid.

Bertrand's compositions are not only impeccably intelligent, but also a reflection on the art of perfumery: in this case, exploring vanilla as though it were a strange new material and deducing its place on the scent-map. The beauty of them is that they also tell a story. Think of vanilla and you're already in Central America, from where the plant originates. From there it's only a short slide to the Caribbean islands and two of their chief luxury exports, cigars and rum, both of which share common facets with the vanilla pod. Again: logical.

But if the fragrance is a thinking woman's (or man's) vanilla because of the new light it sheds on the genre, it's also a sultry, Carmen-rolling-cigars-on-her-thighs scent. Is it useful at this point to mention that the very word 'vanilla' comes from the Latin for 'sheath', *vaina*? Just add that missing letter – the erotic subtext is part of vanilla's appeal. Not to mention that, despite what Freud once quipped, a cigar is not necessarily always a cigar …

This is what's been making Bertrand's work so interesting of late: the feeling that he has been engaging more sensuously with his materials. His scents used to be fascinatingly weird, dark and austere, as though he were making a point of holding at arm's length the more pleasing aspects of perfumery. But his latest stuff has been getting hot and bothered, languorous and dirty; it's growing flesh. He says it's because, since he's set up as an independent, he has a more hands-on relationship with his materials – in his old job, he wrote down his formulas and an assistant blended them – but also because he is now allotted larger budgets and can use higher quantities of the better, richer stuff. Yet I'm not quite sure it's only that. His perfumes still have quirky notes, but they're … more *pleasing*. More wearable.

'More commercial, you mean? Well, maybe now that I'm independent I feel more responsibility …'

That's not what I mean. To me, it's as though at this stage of his career, he doesn't feel as much of a need to go for the weird. As though he can allow himself to play with more outrightly seductive notes without having the feeling he's selling out …

That stumps him a bit. He knits his bushy eyebrows.

'Maybe. I don't know.'

He can afford to stray out of his own weird comfort zone, I tell him. After all, he's one of the best perfumers of his generation. At that, he blushes deeply, cocks his head, mutters the Gallic equivalent of 'Aw, c'mon', and gives a little kick to the tip of my boot while staring at his own. Now it's my turn to squirm on my chair. I'm no Coco Chanel, though I don't like roses much either. I don't care about launching another N°5. But, like Chanel, I know what I want, and now's the time to ask for it.

'So, remember that story I told you about Seville? You said it would make a good perfume …' At that point, I'm loath to confess, I consider leaning forward a bit to flash some cleavage – a tactic which, I've learned during my years as a journalist, quite efficiently throws male interviewees off their stride. It's an urge I curb. '… does that mean you might want to go ahead and make it?'

He pauses, nods and looks me straight in the eye.

'OK. I'm game. Let's do it.'

And that would just about be when I faint.

7

What do Michael Jackson and my mother have in common? They both wore Bal à Versailles, by Jean Desprez. What's shocking about this piece of information is not that Michael Jackson and my mother had a point in common: in fact, they had two since they both married a Beaulieu, Lisa Marie Presley's mother, Priscilla, apparently being a cousin of mine to the nth degree. It's that my mother owned perfume at all. And that the fragrance she picked would be so … pungent. Clearly, when she indulged her rebellious side, she didn't go for half-measures. In the mid-70s, there were a lot of fragrances that would've suited her brisk, no-nonsense personality much better, like Diorella or Chanel N°19. But no: she went for something lush, warm and powdery in the most classic tradition of French perfumery.

I found her secret stash while I was alone in the house, rooting around for her makeup bag. After seven years of frenzied gardening, oil painting and sewing my clothes, hers and my Barbie dolls', my mother had gone back to work as a nurse at the local hospital the instant I'd started high school. A few months shy of my thirteenth birthday, I wasn't allowed makeup and the nuns were strict about enforcing that rule. The grey-lipped Sister

Jeanne would drag us by the arm into her office to blast off the tiniest trace of lipgloss, just as she made us kneel to measure how far up our thighs our miniskirts rose. So, of course, the forbidden pleasure of makeup was all the more covetable.

My mum must have wised up to my covert raids because one day, the makeup bag disappeared from under the bathroom sink. Her nightstand was the obvious place to look, and there it was, as expected, wedged between two books and a box. I'd just stumbled on my mother's secret life.

I'd always had full access to the family bookshelves, but these two particular volumes were understandably not meant for a young girl's eyes: *The Sensuous Woman* by 'J' and *The Female Eunuch* by Germaine Greer. It's a wonder my barely pubescent head didn't explode after reading a sex manual and a feminist essay hot on its heels. There was 'J', explaining 'how to drive a man to ecstasy' with 'the Butterfly Flick' and 'the Silken Swirl' (I practised them on an ice-cream cone), all the better to catch a male and keep him. And there was Greer, thundering that learning to catch a male and keep him, as girls were trained to do from an early age, led straight to the frustration, rage and alienation of the suburban housewife. 'J' advocated games, disguises and elaborate sexual scenarios – the scene in which a wife re-does the connubial bedroom with mirrored ceiling and leopard-print bed-sheets was seared in my mind for ever, and may explain the pattern of the cushions on my couch. 'I'm sick of the masquerade,' raged Greer. 'I'm sick of belying my own intelligence, my own will, my own sex … I'm sick of being a transvestite. I refuse to be a female impersonator. I am a woman, not a castrate.'

Though they couldn't have been more different, both books were pure products of the sexual and feminist revolution of the 60s, and they did have one message in common: women had sexual desires and wanting to fulfil them didn't mean you were 'bad'. I was still a few years away from putting the Silken Twirl into practice, and nowhere near renouncing feminine adornment since I barely had access to it. In fact, I was still young enough to remember the fun of dressing up in my mother's cast-offs and Geneviève's fripperies. But I did learn a lesson from Greer: the trappings of femininity to which I so aspired were just that, trappings. You could put them on and take them off. You didn't have to *be* them. It could be a game, like playing dressing-up, rather than an obligation. Somehow, the rich, ripe Bal à Versailles I kept sniffing while I read the forbidden books became enmeshed with those lessons. Perfume, at least in my household, was as subversive as sex or feminism: a claim laid to a world beyond Ivory Soap and lawnmowers, as well as my mother's personal manifesto against bedsores and bedpans. At the Lakeshore General Hospital, she dealt daily with the human body in its most inglorious states. Nature she knew, and she was doing her best to stop it from doing its stinking, miserable worst. But she also wanted an option on artifice; she wanted Versailles and the useless, futile, sinful beauty of it; she wanted to go to the ball.

My father was far from alone in his indictment of perfume: though his aversion was probably due to his hyper-sensitivity rather than to moral or philosophical reasons, he was in fact only following in the footsteps of a long line of male authority figures. Philosophers, priests and physicians had been railing for centuries against it, cursing women for snatching it from the altars of the gods and diverting it for erotic purposes.

The Ancient Greeks were the first to sound the alarm on fragrant substance abuse. Perfume was the potent lure used by the enchantress Circe to ensnare the wily Ulysses while she turned his companions into swine, and by the courtesans who diverted the fruitful union of man and woman in their sterile embraces ... A measured, reasonable use of scent could stimulate desire, the wise men granted, hence its use in wedding rituals, but the operational word was 'measured'. In *The Republic*, Plato warns against 'clouds of incense and perfumes and garlands and wines, and all the pleasures of a dissolute life' that turn 'the sting of desire' into madness. Man must not be led by lust and, though a drop of sweet-smelling stuff might promote marital harmony, an excessive use of it disrupted the balance between mind and body. Though the very nature of perfume was divine because all beauty springs from the divine, explains the historian Jean-Pierre Vernant in *The Gardens of Adonis*, 'in erotic seduction it is diverted, led astray, directed towards a pretence of the divine, a misleading appearance of beauty concealing a very different reality: feminine bestiality.'

But it was for its wastefulness rather than its immorality that the hard-nosed Roman empire condemned it. 'Perfumes form the objects of a luxury which may be looked upon as being the most superfluous of any, for pearls and jewels, after all, do pass to a man's representative, and garments have some durability, but unguents lose their odour in an instant, and die away the very hour they are used,' grumbles Pliny the Elder in his *Natural History*. 'The very highest recommendation of them is, that when a female passes by, the odour which proceeds from her may possibly attract the attention of those even who till then are intent upon something else.' Talk about damning with faint praise.

As for the Old Testament, it is lush with fragrance – 'My lover is to me a sachet of myrrh resting between my breasts,' says the 'dark but comely' bride of the Song of Songs. And the New Testament does feature two fragrant episodes: in the first, an unnamed sinner washes Jesus' feet with her tears before anointing them with precious oils. In the second, it is Mary of Bethany who pours costly spikenard on his head in anticipation of his funeral rites. Even then, there's male grumbling, since Judas considers the money would have been better spent on helping the poor. But as a rule, Christianity dealt much more harshly with scents. They were part of pagan rituals, especially the highly popular imported Asian religions which competed with the new Christian cult in the Late empire; worse still, they could induce men to sin. Not only did Christian women need to distinguish themselves from the pagans who marinated themselves in scent, but they should strive to make themselves ugly so they wouldn't arouse lust, urged Tertullian in the 2nd century. Woman was 'the gateway of the Devil' and by adorning herself, she not only led fellow Christians astray: she defaced the work of God. The body was contemptible and so were bodily pleasures. The only sweet smell was that of a spotless soul. Sin stank.

For the Church, saints were the only beings whose corpses didn't exhale the stench of corruption which was the destiny of all living creatures. Their mortal remains were said to give off suave effluvia years or even centuries after their death: the odour of sanctity. Conversely, the strongest stench was raised by the ultimate sinner, the heiress of the temptress Eve whose wiles caused humanity to be cast from the Garden of Eden: the whore.

The etymology of many Latin language words for 'whore' (*pute* in French) is derived from *putere*, 'to stink'. Prostitutes, it was believed, wore fragrance both to attract their prey and to cover up the emanations of the male secretions festering inside their bodies, venom they spread from lover to lover. When the olfactory-obsessed Émile Zola writes about the courtesan Nana in the eponymous book, he calls her 'that Golden Creature, blind as brute force, whose very odour ruined the world', even though, when he does mention her favourite scent, it is the mawkish violet. But her streetwalking counterparts would have wafted headier aromas such as musk or patchouli as olfactory advertisement for their wares.

The association between the putrid *puta* and her fragrance abuse is embedded in the subconscious perception of perfume, but in 1932, the owner of the Spanish perfume house of Dana went straight to the point when he asked Jean Carles to compose a *perfume de puta*, a 'whore's perfume' – surely the raunchiest brief in history. He called the orange blossom, carnation and patchouli blend Tabu, the 'forbidden' perfume, after Freud's *Totem and Taboo*.

Think of it the next time you spritz on a juice called Obsession, Addict, Poison or Aromatics Elixir. You're not just doing it to smell good: you're perpetuating a ritual of erotic magic that's been scaring and enticing men in equal measure for millennia.

8

It wasn't about the smell back when I was a teenager. It was about the ads.

In my all-girl Catholic private school, my classmates had nicknamed me 'The Dictionary' because I used words they didn't understand. I was an outcast, and magazines were my only access to the stuff they talked about at break, my only clue to becoming a woman. Worse still, my body had declared war on me, growing tall and sprouting fat so quickly I was striped with purple stretch-mark welts all over, and perpetually falling flat on my face because my centre of gravity kept changing location. Not only was I a geek, but I'd become a chubby, bespectacled geek. Fourteen sucked.

My best friend Sylvie was ostracized for the opposite reasons. The other girls called her '*La Guidoune*', a Québécois slang term for slut. They sniggered at the blowsy D-cup breasts that tugged her blouses open on her greying bra, her rats'-nest hair, occasional bouts of funkiness after gym, and the way she sprawled behind her desk, knees and lips parted, staring at her chipped nail polish. But I envied her the way her breasts rolled and bobbed under her nylon blouses, the boys from the vocational school who hung around a block down to pick her up after class, and even the sovereign vacancy with which she greeted anything that didn't have to do with beautification. My own mind was a jumble of the things I'd read and was trying to make sense of; my only-child life was boyless, since my freckled next-door neighbour Jeffrey was a hockey-obsessed jock and Jacob, two doors down, was only willing to bond over his pet iguana: I was the sole girl on the block who didn't run away shrieking when he slid its head into his mouth.

Lunch breaks with Sylvie meant greasy brown vinegar-doused chips bought from a trailer and lengthy browses in Woolworth's cosmetics aisles. In a burst of teenage rebelliousness, I'd decided to transgress the paternal ban and buy my first bottle of perfume with my babysitting money, a purchase discussed at length with my best friend during break. I'd whittled down my options to three possibilities after studying the ads. The one for Revlon's Charlie, 'The Gorgeous, Sexy-Young Fragrance', featured a grinning model striding confidently in a trouser suit. I rather fancied being a freewheeling career woman with legs a mile long, but Sylvie pulled a face.

'She looks like that stupid guy on the Johnny Walker bottle.'

With its faux-fur cap, Tigress by Fabergé spoke to my worship of all things feline and reminded me of *The Sensuous Woman*'s advice on leopard skin-patterned sheets. The ad intrigued me: a gorgeous black woman on all fours wearing a tiger-print body suit and a slight smirk. 'Tigress. Because men are such animals.'

'Uh-uh. Get that one.'

Sylvie pointed a frosty-pink nail towards 'Love's Baby Soft. Because innocence is sexier than you think.' Today the ad, a pouting girl clutching a white teddy bear while fully made-up and coiffed though she couldn't be more than twelve, would have child-protection leagues tear down the offices of Menley & James Laboratories brick by brick. Even back then I found it creepy. I was innocent in body if not in mind: 'J' had made sure of that, and the theoretical knowledge she'd imparted fuelled my reveries of teenybopper idols like *The Partridge Family*'s David Cassidy. Along with Jovan's Musk, Love's Baby Soft was the most popular fragrance at school and perhaps the key to some measure of acceptance. As Sylvie wandered off to check out the Revlon display, I pondered buying into what was, in effect, the tribal smell of middle-class teenage girls in the 70s – sweet, powdery, faintly sickly – and wondered just what it was about 'musk' they all considered so 'sexy'. Perhaps I was too ignorant, despite 'J's best efforts, to know what sexy was. Little did I know I'd stumbled on one of the greatest paradoxes of perfume ...

In its natural form, extracted from the abdominal pouch of a species of deer native to the Himalaya named the musk deer, musk smells of honey, tobacco, fur, earth, man, beast ... But in its various synthetic guises, it whispers of just-out-of-the-shower freshness, clean laundry and powdered baby bottoms. To sum up: the same word means both 'clean' and 'dirty', 'innocent' and 'sexy'. Of course, clean smells may arouse dirty thoughts ... Which was, I suppose, the whole point of Love's Baby Soft, but a paradox indeed, and one that sprung from centuries of musk madness.

Musk was so popular in Imperial Rome in the 4th century that Saint Jerome had to prohibit his flock from wearing it, which is how we know its use goes back 1,500 years in the West: China had certainly known it for much longer. Moslems thought its scent so divine they incorporated musk pods into the mortar of their mosques so that, once warmed, the walls would exhale their sweet effluvia. Europe rediscovered musk with the Crusades and Marco Polo's reports from Kublai Khan's empire. It remained popular until the mid-18th century, then enjoyed a brief revival after the French Revolution: in a counter-reaction against Robespierre's bloody reign of Virtue, the Royalist Muscadins adopted it as an olfactory emblem – perfume could literally make you lose your head back then. It fell out of favour once more in the 19th century, when it was accused of causing hysteria, but also used to treat 'sexual torpor' in women.

However, musk didn't disappear from perfumers' armamentaria in the Victorian era. As Septimus Piesse explains in his 1857 *Art of Perfumery*, 'It is a fashion of the present day for people to say "that they do not like musk" but, nevertheless, from great experience in one of the largest manufacturing perfumatories in Europe, we are of the opinion that the public taste for musk is as great as any perfumer desires. Those substances containing it always take the preference in ready sale – so long as the vendor takes care to assure his customer "that there is no musk in it".'

Cheaper synthetics gradually replaced natural musk; in 1979, the musk deer became a protected species, though the use of musk tincture from remaining stocks is not prohibited. Since it's always been costly, labs have been coming up with substitutes for over a century. None has quite managed to replicate the unique properties of the real thing.

It is because Western fragrance companies started delocalizing the polluting production of some types of synthetic musks to developing countries like India that the fashion for musk reappeared in the West after a two-century eclipse. Hippies may have trekked all the way to the foothills of the Himalaya where the musk deer was poached, but what caught their fancy was entirely man-made. And though the suave odour may have blended well with the effluvia of the Flower Children, it had in fact become the olfactory symbol of cleanliness in the West since the 1950s, when synthetic musks started being added to detergents because they remain stable in harsh environments. Generations have come to identify their smell with freshly washed linen, which is certainly why the so-called 'white musk' note has segued so easily from functional to fine perfumery. Perfumers love it too: molecules with such science-fiction names as Galaxolide, Nirvanolide, Serenolide, Cosmone, Astrotone, are capable of boosting other notes, covering up gaps in wonky formulas, expanding their volume and giving them the half-life of plutonium on skin.

I wasn't any crazier about musk as a teenager than I am now and, in the end, I opted for Tigress in the hope that I would find out some day what it was that turned men into animals … Since my own bottle disappeared decades ago, I've asked an American friend to decant a few drops for me from her own vintage stash. I'm on my way to Bertrand's lab when I retrieve a padded envelope containing her sample from my mailbox. And so it is with him that I catch my first whiff of Tigress in over thirty years.

'This is what you wore when you were fourteen?' Bertrand grins.

'I remember *buying* it because of the ad …'

We Google it and he lets out a hoot.

'I can see why! And did you wear it?'

I sniff the blotter. No memories of slumber parties with Sylvie or flashbacks to sniggering classmates aping my 'French' accent pop up. All that Tigress brings to mind are the other perfumes its carnation, sweet balms and cheap rose notes remind me of. The masterful 1912 L'Origan by Coty – an olfactory punch in the gut the first time I smelled it four years ago; Tabu by Dana, which had become drugstore swill by the time I acquired my Tigress; Estée Lauder's iconic Youth Dew, whose spicy facets permeated the ground floor of the local department store of my youth; its descendent Opium by Yves Saint Laurent. I could rebuild just about half the history of perfumery from those few drops of Tigress, but not my teenage years. My olfactory culture seems to have crushed any earlier memories. Have I actually worn this? I'm starting to doubt I even owned it.

Besides, Tigress isn't the reason that brought me to the lab today. I've come because Bertrand has composed what might become the core of our perfume.

'So … You said you had something to show me?'

* * *

'These aren't perfumes yet.'

I nod. What Bertrand is showing me today are two sketches for the scent built around orange blossom and an incense we've code-named 'Séville Semaine Sainte'.

I've deliberately refrained from trying to envision what I'm about to discover, or even from recalling the smells I experienced in Seville. I don't know what to expect, or what's expected of me. I've never done this before: never been in on the very first steps of the conception of a fragrance, much less one inspired by me. In exchange for my idea, I'll be following its development, recording our sessions and writing a journal of our creative journey; I'm also to keep a sample of every version of the formula. How far Bertrand will take the project, I can't fathom. So far he's walked his talk: said the story would make a good perfume, said he'd make the perfume, is now making it. For the time being, this is a purely personal undertaking on his part. Of course I can't help wondering whether it'll ever come out, but that's not how things work: perfumers don't just waltz into a client's office brandishing a finished product. If the scent comes to term without being marketed, Bertrand will have made me a gift worth several thousand euros, the cost of developing a bespoke perfume for a private client. But this isn't a bespoke perfume, is it? I'm not asking for an olfactory mirror. I just wanted to walk through the looking glass. And I'm about to.

Bertrand labels his blotters '1' and '2' and dips them into the phials.

'You'll find the first one is a lot more austere than the other.'

I breathe in. This is soapy. I can pick out incense ... Aldehydes with their characteristic snuffed-candle and citrus ... Lavender ... Eww, I hate lavender ... But I can't really detect the orange blossom, though Bertrand says he's boosted its tarry notes with yara-yara, the material I discovered here a while back that reminded me of the medicinal effluvia of my childhood, and indole, a mothball-smelling molecule found in flowers like jasmine, orange blossom, honeysuckle or narcissi.

'You've got to imagine a street in Seville baked by the sun, with the tar almost melting, just before a storm,' he explains.

'You got that from the story?'

'Absolutely.'

I'm a little disconcerted, not only because I don't remember mentioning any melting tar, much less storms, but because with its soapy lavender notes, I find N°1 jarringly masculine. N°2 is brighter, almost tart, but also softer and more suave. The orange blossom absolute has been fleshed out with jasmine and aurantiol, a base resulting from the reaction of methyl anthranilate (the main odorant molecule of orange blossom) and hydroxycitronellal (which smells of lily-of-the-valley).

But there's something else lurking beneath the sweetness. Something a little ... beastly? The afternoon is uncharacteristically hot and muggy for the end of April and my body feels slightly damp under my black cotton shirtdress. Has my deodorant let me down?

'Excuse me, I'm about to do something not very lady-like ...'

I lift my collar away from my shoulder and take a quick sniff. Bertrand lets out a mischievous little giggle, like a kid who's played a neat trick on the grown-ups.

'I've put in some costus, to give skin and female scalp effects.'

Costus smells of fur and dirty hair, he explains: force the dose, and you'll get badly cured sheepskin. Incense is also tricky to work with, since it can produce repulsive facets of raw flesh and butcher's stall. So he can't use a high percentage of it and has to boost it with other materials that have incense-like facets, like aldehydes and pink pepper.

I must look a little disappointed – I guess I somehow expected the magic to work straight off. After a moment's silence, Bertrand speaks up:

'These are just first drafts. They're still pretty austere for the moment. I've done them to find out where we set the cursor. But we haven't necessarily found the accord yet.'

N°1 is definitely too soapy, I tell him. There *are* soap notes in the story wafting from the crowd, but they should be fleeting impressions. And N°2 is too sunny. In my story, the only light comes from the candles flickering on the gold of the float.

Bertrand frowns, clearly trying to figure out how this translates into olfactory terms. We've known each other for nearly five months now, we've talked for hours, but this is a new type of conversation and we need to adjust our languages.

'You mean it's too floral?'

Well, no, the scent needs to be floral because there *are* a lot of flowers in this story, with all the lilies spilling out of the float, I venture.

As Bertrand stifles a sigh, I realize I've just steered him in a new direction.

'Would you rather go for a lily than an orange blossom?'

Instead of answering, I blurt out:

'And there are tons of carnations too …'

Now I've done it again, haven't I? But he nods patiently.

'Right. Carnation. That's very spicy. For the moment, I'm not very spicy. I'm indolic.'

Eugenols, the molecules that produce the clove-like smell of carnations, belong to the same chemical class as indoles and phenols, Bertrand explains. But they 'vibrate' in different ways: eugenols burn, while indole and yara-yara melt, 'like tar in the sun'.

As soon as he mentions melting, I'm reminded of beeswax. During the Holy Week, little kids collect it from the penitents, who tilt their candles so that a few drops will fall on the children's wax balls.

'OK, we'll put in beeswax … But all those things are very *austere*, you know? If that's what I do for you, it'll be as dark as the darkest night. You'll barely be able to make out the gold. We've got to find the night lights.'

He's right. This shouldn't be austere. I'm in the arms of a boy who's got a hand under my skirt. That's what makes the meeting of incense and orange blossom so symbolic, this blend of the sacred and the erotic …

'It'll be tough to do something pleasant,' says Bertrand, 'because orange blossom and incense are two *hard* notes. If you want to make them prominent, you're going for hard on top of hard.'

But the notes shouldn't be a pretext, he adds, otherwise there's no point. You can't say you're doing an orange blossom and incense fragrance then stick in a couple of drops just so you won't be an outright fraud, like most perfume companies do nowadays.

I can't help feeling a little smug. I've presented him with a challenge, and I'm starting to know him well enough to understand he thrives on challenges. So I try to help him the only way I know how, by telling him more about Holy Week, hoping that among my words he'll find something that teases at his own memories, that translates into his own language; something that'll make this scent as sensuous and seductive as Seville abandoning itself to the religious-pagan fiesta. The exhilaration of a city flowing from street to plaza to get a glimpse of the processions; the bar-hopping instead of the Stations of the Cross, the sea-salt aroma of the blond *vino de manzanilla* and the bittersweet herbal pungency of joints; the dizziness and flirting and laughter. The dark, thrumming beat of the drums, the solar jarring bursts of the brass bands, the beeswax coating the streets with a silky sheen, feet slipping as the crowd mills about. The darkened plaza where the float appears, blazing like the ocean liner in Fellini's *Amarcord*, with the musty whiffs of derelict palaces seeping through shutter windows behind the wrought-iron grilles …

It's a strange sensation. This man is so open, so willing to be enthralled, that I get the feeling he's with me in the jostling crowd.

'Fascinating. This isn't my world at all, but it could've been. I must've lived this before, in another life, because it speaks to me so much.'

'It's as though I were trying to draw you into my memory.'

'But I *am* there. Completely.'

9

Was that guy following me?

I tried walking faster but my strappy sandals made my steps wobbly and my 40s wraparound dress kept flapping open: I had to slow down to press it shut on my thighs. Paris is practically empty in August as Parisians migrate en masse to the beach, but there they were, the summer bachelors in their suits and ties, wife and children packed away, wandering out of their offices and picking up my trail. Appraising looks at a silhouette that was starting to shape up … Slowing down in front of shop windows when I looked over my shoulder … This was the first time I'd wandered off into Paris without my parents, and it felt as though the whole city was hounding me.

We'd finished paying off the house and the dollar was high, so we could afford our first European holiday. Being French-Canadians, our destination was never even discussed: it would be the old country. Paris, the very place I'd so aspired to come to a mere six years ago. But at seventeen, you didn't *do* excited when you were trailing behind your parents, wishing you didn't look as though you were with them. This was just a scouting expedition. I'd be back on my own some day. By then I might have figured out how to handle the attentions of the older males of the Parisian herd. Who did those guys think I was?

Sweat was rolling between my breasts. My stalker was still there, some guy idling away his lunch hour, probably enjoying my panic, his hunter's instincts aroused. I ducked into a perfume shop – the touristic Avenue de l'Opéra hadn't been hit by the annual holiday shut-down. A slight, dark woman with side-swept hair framing a heart-shaped face, her crimson lips a vivid contrast with the lapis lazuli of her drawstring-waist dress, gave me what I would later come to call 'the Parisian bar-code gaze': a jaded, swift, head-to-toe assessment of market value. She made me feel as though I'd crawled out of a trashcan.

'May I help you, mademoiselle?'

I peered through the window. My stalker was leaning down to spy on me between the Christian Dior displays.

'Actually, I came in because this man was following me … No, don't look! Can I stay here for a while?'

I was beginning to amuse her. Slightly.

'Stay as long as you want.'

I fussed with the lipsticks, jabbering about annoying Parisian men. Lapis Lazuli cocked her head on her shoulder.

'And why does that bother you?'

'I find it … *insulting*. As though they thought I was cheap.'

'They're just trying their luck. No harm done. Let me know if I can help you with anything.'

I turned my attention to the bottle display. Since Tigress, I'd managed to smuggle Shalimar talcum powder and a Coty Sweet Earth compact with three small pans of wax smelling of hyacinth, honeysuckle and ylang-ylang into the house. The latter's scent stayed fairly close to the skin and I'd found I could risk it even in the vicinity of my father's oversensitive nose. But the stuff on those shelves was better: a dive into the glossy pages of French magazines.

'That's a very pretty dress you've got there, mademoiselle. Did you find it at the flea market in Saint-Ouen?'

I'd been to Saint-Ouen, of course, though the dress came from a vintage clothing stall in downtown Montreal. Since I'd stumbled on it, my closet had become crowded with what my mother disgustedly called 'old clothes' – 40s crepe dresses, ruched silk blouses and nip-waisted, shoulder-padded satin jackets. Clearly, I was on to something if this elegant Parisian thought my dress was pretty …

I reached out my hand to a bottle shaped like a flattened drop, took off the cap and was on the verge of raising it to my nose when Lapis Lazuli snatched it, deftly dipped a blotter into it and waved it around in the air before handing it over.

'*Voilà*, that's how it's done,' she said in that dry, pedagogical voice Parisian women use when they're handling clueless tourists. Her bronze-lacquered eyelids fluttered as she brushed me with another bar-code look, a little more slowly this time.

'You're lovely … so *pulpeuse*.'

It was the first time I'd heard that particular French word, a synonym for plump, perhaps, but one conjuring ripe-fleshed fruit rather than blubber. Lapis Lazuli had pronounced it as though she'd kissed the air twice. I wondered whether she was coming on to me. In fact, she was closing in for the kill.

'You need a luscious, ample, floral scent to suit those naughty eyes of yours peeping under your fringe – no wonder men are following you around … This is First, by Van Cleef and Arpels. I've always thought jasmine was wonderful with pale olive skin like yours …'

Pearly bubbles popped inside my nose while I flailed about for my scant olfactory references.

'This reminds me a little bit of … Rive Gauche, maybe?'

Lapis Lazuli nodded.

'It's in the same family, but it's more opulent. It may be too ladylike for you though. How about Azzaro? It's also got jasmine, and rose, and gardenia, but with a warm amber and sandalwood base. The little touch of plum gives it a lovely rounded feeling …'

I'd never heard anyone speak about fragrance that way. Apart from my dissections of ads with Sylvie to pick out the women we'd become if we sprayed on the potions, I'd never discussed fragrance at all. Of all the notes Lapis Lazuli had mentioned, only rose and plum came within the scope of my experience. But Azzaro did feel like a better fit than First. There was something tender about it, an earthiness seeping through the flowers.

I wandered over to a round bottle with a cap in the shape of twin stylized arums. This brand I knew better. In a bid to wean me off my 'smelly old rags' my mother had sewn me a flowered chiffon drop-waist dress by Chloé from a Vogue Paris Original pattern and I loved the way the drawstring neck slid off my shoulder when I loosened it after leaving the house.

'Why yes! Chloé. Why didn't I think of it myself?'

Smothered in an avalanche of white petals, I gasped, inhaled again, and let out a little delighted chuckle. This I could understand: FLOWERS.

'I'll take it!'

Lapis Lazuli tut-tutted.

'You need to know if it likes your skin and if your skin likes it. Try it on for a while.'

What, she didn't want to make the sale now?

'Try Chloé on one wrist and Azzaro on the other. You can come back later.'

I did come back the next day for a bottle of Chloé. Not because I was particularly smitten with it: for all her frothy blondeness and floaty flowered chiffon, Karl Lagerfeld's first fragrance felt like she could pogo me off the dancefloor without breaking a stiletto heel. But it would have been hard to find a scent more obnoxious to my father; the olfactory equivalent of the Dior scarlet lipstick I also bought from Lapis Lazuli when I went back.

Needless to say, that huge wake of flowers drew even more Parisian flesh-hounds. Some of Lapis Lazuli's sexual self-confidence must have seeped under my skin, because I was starting to feel it was fun. I'd even dawdle and throw backwards glances from under my fringe. My body may not have had a *Vogue* model's tawny long-limbed elegance or Lapis Lazuli's dusky wiriness, but it had magnetic power. Men *were* such animals …

My Parisian initiation came at a turning point for the industry. The 1976 First by Van Cleef and Arpels was actually 'the last major perfume of this century which was developed in the classical manner, the last perfume not to use marketing', its author Jean-Claude Ellena told Michael Edwards in *Perfume Legends*. Later on, in his *Journal of a Perfumer*, he added that he had 'collected, borrowed and piled up every sign of femininity, wealth, power' to compose it.

The following year, a perfume was launched that outdated the bourgeois charms of the Van Cleef and sent perfume executives scurrying to huddle with their marketing teams; a perfume that tapped directly into the subconscious perception of perfume as a mysterious, intoxicating substance that could magically transform women into exotic empresses.

Yves Saint Laurent's 1977 offering sprung, as fragrances had since Poiret's day, from the sovereign decision of a couturier. What was unprecedented was the strength and consistency of its scandalous concept.

Opium wasn't the first perfume to be named after a drug. The 1933 Cocaina en Flor, by the Spanish house of Parera, was promoted as 'a mysterious perfume … which enthrals, attracts, bewitches'. Further back, the industrial-strength 1927 Russian perfume Krasnyi Mak, 'Red Poppy', overtly played on balsamic, spicy notes evocative of opium resin, a product of Soviet Afghanistan. Did the same concept inspire similar olfactory results? 'Red Poppy' smells like a forerunner to Opium. But the drug that Yves Saint Laurent was thinking of in his blissed-out Marrakech retreat was LSD, according to the designer Pierre Dinand, who went to meet the couturier in his Villa Majorelle to discuss the bottle. The name Opium sprung from the Japanese *inro* Dinand brought as an inspiration in a later session. Yves Saint Laurent recognized it instantly as the phial samurais used to carry their salt, spices and … 'opium!' he exclaimed. The American pharmaceutical company that owned the licence for Yves Saint Laurent perfumes, Squibb-Myer, was understandably horrified, but Saint Laurent was adamant – and right. The scandalous, exotic, mysterious Opium was appropriated by women all over the world precisely because, like the drug, it promised escape. The perfume itself was of unheard-of strength and concentration for a French fragrance. Up to then, only Estée Lauder had offered such powerful brews and, in fact, the American grand dame was distinctly displeased at recognizing a formula quite similar to her own Youth Dew (she retaliated with Cinnabar, in what would be dubbed 'the war of the tassels'). Industry insiders also say Opium was actually quite cheap to make, which would induce a downward spiral in budgets.

Opium was Yves Saint Laurent's swan song as 'the designer who gave power to women', in the oft-quoted words of his partner, Pierre Bergé. Despite its transgressive allure, it expressed the couturier's retreat from the Rive Gauche and the liberated Parisiennes he had induced to wear trousers into the exotic, colour-saturated dream worlds he'd been exploring in Marrakech. The woman 'addicted to Yves Saint Laurent' embodied by Jerry Hall in the first print ads had dropped the keys to power to lean back languidly on the gold-embroidered cushions of her Chinese den, where she rested after wild nights at Studio 54 or Le Palace. Or was it that, sated by Opium, the supine Amazon had at last shed her need for men in an ultimate declaration of independence?

Breaking away from the real, albeit privileged, world to set off for imaginary lands, couture became a spectacle. Over the following years, its relevance waned; couturiers ceased to dictate styles and so, essentially, became image purveyors driving the sales of cosmetics, perfumes and accessories. The conception of designer fragrances would be taken over by the marketing departments.

The launch of Opium in North America was also a watershed moment in my olfactory life. I'd come back to Canada after my trip to Paris determined to save up enough money to go back to France to study. Over the Christmas holidays, I worked as a gift-wrapper in a high-end Montreal department store, right next to the perfume counter. This was where I contracted a lasting addiction to cashmere – I had to fold those sweaters before wrapping them up – and a Pavlovian loathing for Opium, which the sharp-taloned Hungarian Lisa would spray all day and sell by the gallon to last-minute male shoppers. My stint as a gift-wrapper, which went on for three years, practically vaccinated me against the whole of the Estée Lauder opus up to 1980 and most of the better-selling classics – N°5, Arpège, L'Air du Temps – whose every waft was tainted with Opium. The potent mix clung to my clothes; it became associated with Lisa's hypnotic sales pitches, my aching feet and another type of ache, for the beautiful things that passed briefly through my hands and that I couldn't afford. The experience nearly put me off mainstream fragrances altogether, at an age when most teenagers were scrimping to buy their first bottle of Anaïs Anaïs or Cristalle.

Today, standing in the cavernous Sephora flagship store on the Champs-Élysées, buffeted by waddling bum-bagged tourists and fleet young black-clad sales assistants, I wonder how I'd go about choosing my first grown-up perfume. The wall of fragrances must cover the better part of a kilometre; atomizer-wielding demonstrators lie in ambush and avoiding their spritzes requires ninja-like skills. The conversation I had with that glamorous Parisian shop manager back in the late 70s I could never have here. And the fragrances I was offered then, opulent stuff with breasts and hips and a regal stride, gather dust on the bottom shelves, if they've survived at all (the original Chloé hasn't). Teenage girls are the target demographic for practically every mainstream launch; brands fall all over themselves to cater to their tastes. The Max Factor Green Apple that felt like a slap in the face to me has now grown into a fruit basket the size of the Himalaya and spilled out into every shopping mall.

The cheery, unsophisticated berry had been bumping for decades at the door of perfumers' labs before someone wondered what that squishy noise was, saw a lick of red juice trickle in, opened up and … Ker-plash! The whole crop spilled into the vats. Soon, even legendary perfume houses such as Guerlain were plonking the notes into the mix: perfume had suddenly gone *pink*. The berry binge introduced within the codes of fine fragrance a type of note that had come up from functional perfumery. It used to be the other way round: if a perfume was popular, functional fragrances copied it in a simpler, cheaper form. This is why many older brands of hairspray smell of Chanel N°5 or L'Air du Temps; why shampoos in the 70s had the green notes made popular by Chanel N°19 or Givenchy III. Why, at least five products in my bathroom smell of L'Eau d'Issey at this very minute. But in the 90s, the notes of functional fragrances started trickling *up* into fine fragrance with the synthetic musks used in detergents when the public started craving 'clean' in a reaction to the over-saturated scents of the 80s. Perhaps not so coincidentally, this happened at a time when detergent companies were busy buying up perfume houses, their executives smoothly segueing from washing powders to fine fragrance. Fruity notes were the second wave in the 90s, a trend for which the American consultant Ann Gottlieb claimed responsibility. When hired by Bath & Body Works, she said, she introduced 'things that, up until then, women had found almost nauseating. These fruity notes then came into the public domain much more, and people started loving [them].'

OK, so now those of us who *still* find them nauseating know who to blame for the wall-o-fruit we crash into as soon as we step into the mall. While I can only congratulate Ms Gottlieb on her success and influence, I can't quite find it in me to be grateful to her. Especially since her claim demonstrates that the trend sprung from massive clobbering rather than public demand. Granted, the public may have a yen for cheerfully regressive, synthetic scents that remind them of boiled sweets or shampoo – familiar and easy to understand in our sound-bite, mouse-click, twittering ADD world. In a way, Love's Baby Soft is *still* what little-girl scents are made of; spayed smells for female eunuchs. If I were sixteen today, what would I do? Probably pick the latest from a brand I liked. Empty the bottle. Then switch to something else. Would it even be possible to feel the fierce commitment I felt for the first fragrance I truly made my own?

Of course, it helped that I'd fallen in love.

10

Onscreen, Fred MacMurray was ringing the doorbell of a Spanish-style Los Angeles house. In a minute, he'd be leering at Barbara Stanwyck's anklet. In half an hour, they'd be plotting to bump off her husband. I'd seen *Double Indemnity* before. My eyes wandered from the screen to the silhouettes in the first row, bathed in silvery light. Since Concordia University's film noir retrospective had begun, I'd been sitting behind them: the tall, quick, witty Michael, slim as a brushstroke of Indian ink in his sharp-shouldered Thierry Mugler suit; Jon, his scruffy friend in the scuffed aviator jacket, with his stubborn jaw, knobby wrists, light-brown curls tumbling on his forehead; Lise, a poised, slant-eyed blonde with a whispery voice, cinched waist and early-60s pumps; and Mimi, a petite, sarcastic brunette with scarlet lips and schoolteacher cat's-eye glasses. I'd been breathing in the Waft. I couldn't make out which one of them wore the bitter leather and ashtray fragrance that rose up from the first row they'd commandeered. They all seemed to trail that after-hours cloud. Once, I'd lingered in the auditorium after they'd left – we'd got as far as small nods and half-smiles – and leaned down on the scruffy one's seat: the still-warm fabric had soaked up the scent. It felt as tough and dark and raspy as Barbara Stanwyck's voice. I had a crush on all of those kids, but a little bit more on Jon.

'I don't need to use up my bottle of Van Cleef. I'll just sit next to Denyse here.'

Michael plonked himself down on the couch next to me, comically fanning himself with his hands. He was the little gang's charismatic leader, fuelling our discussion with esoteric references to the Bauhaus, Russian Constructivists, Beat poets and Tamla Motown . . . This was the first time I wore their Van Cleef and though I'd felt a little self-conscious about appropriating the Waft – they *did* all wear it, boys and girls, as it turned out – I'd pretty much spray-painted myself with it in Jon's bathroom.

Jon had been the first to speak to me. After the film, he'd invited me back to his flat, where they all hung out before hitting a gay disco in downtown Montreal – they'd crash into the DJ's booth to pester him into playing the selection they felt like dancing to that night. By then, I'd become sufficiently adept at manipulating style to impress even this bunch of sartorial semioticians: punk rock had been a liberating experience to which I'd applied my head-of-class analytical skills. Punk meant you could be a scowling mortadella trussed in dayglo fishnet stockings and still be light years cooler than any Farrah Fawcett blow-dried clone. It was all about *playing* with signs of the ugly, the shocking, the rejects of mainstream codes. Bathed in the Waft, I knew I was finally in with the in crowd, art-school post-punks who revived styles at such an accelerated pace we lurched from 60s bubblegum pop and *Star Trek* kitsch to Beatniks and free jazz within a single summer.

While I still lived at my parents', my black bevelled bottle of Van Cleef stayed stashed in Jon's bathroom to circumvent the paternal ban. The smell of it on my clothing was a way for me to linger in Jon's aura after I'd gone back home to the suburbs, and then, after that summer, to my campus room. Being eighteen with a hopeless crush on my best friend was a dull, delectable pain I sharpened by wallowing in his smells. Every scrap of the Van Cleef carried a bit of him and of our time together. The bitter herbal aroma of the joints we'd puff on while discussing Adolf Loos' *Ornament and Crime* before checking out the local bands. The whiff of soap on his neck when he shaved, as I was leaning next to him over his sink to paint my face on after he'd art-directed my evening's get-up. The cigarette smoke that lingered in his clothes and hair. The weathered leather jacket I'd snuggle up against as we tottered out of a club at 3 a.m. to have potato latkes at Ben's Deli. The funky, dark, animal waft of his sheets when I woke up on the box spring of his bed – he was crashing out on the mattress he'd pulled on to the floor. By that time I'd graduated from writing the music column for the college paper to freelancing with the two local rock magazines: the older editors were still into the likes of progressive rock, heavy metal or the local Quebec music scene, so I covered all the punk bands that came to Montreal. I sometimes spent the night with one of the musicians I'd interviewed, kids barely older than I on their first foreign tour sharing rooms with their roadies. Then Jon sulked, but not much.

Though it was meant for men and worn by all my friends, Van Cleef and Arpels was the first scent I truly felt was mine. It marked my belonging to a tribe at last. It marked my belonging to Jon: made me his, and made him mine, and made me *him*, more than sex would ever have done. It also marked my final emancipation from the belief in femininity. Up to then, there'd always been a girl who had *It* more than me, that elusive quality of really-being-a-woman – Geneviève, Sylvie or the manager of that perfume shop on the Avenue de l'Opéra. Perfume had been the potion that had promised it could transform me into that woman. Now, like those outfits so camp they could actually get the curvy teenage girl I was mistaken for a cross-dresser – I'd drawn the ultimate consequences of the teachings of *The Female Eunuch* – wearing the Van Cleef finally drove home the lesson I'd learned from Germaine Greer as a pre-teen: masculinity and femininity, as opposed to being a man, a woman or any combination thereof, were just a matter of *signs*. And signs could be played with, believed in just enough to derive pleasure from them. They weren't an identity to be caught up in, yet never felt adequate to. The Waft was the ultimate emblem of my transgression: provocative, invasive, but invisible. People would see a girl and look for the guy who must be lurking behind her. During those heady days in Montreal, fierce in style and intellect, set loose by the crashing chords of punk, I discovered I could be both.

When I sought out Michael thirty years later to ask him how it was that the Van Cleef came into our lives, he answered that he'd been the one to introduce it. It was his return to male fragrances after wearing a mix of Bal à Versailles and 4711, he said, and you couldn't invent that: the boy who introduced me to olfactory gender-bending wore my mother's secret perfume …

Van Cleef and Arpels stretched the spectrum of my olfactory tastes from the shameless floral femininity of Chloé to the toughness of leather and tobacco, which is probably why, twenty years later, I'd slip so pleasurably into the work of one of the first female perfumers, and one of the ballsiest of either gender. Germaine Cellier had already straddled that divide back in the 40s.

With Fracas, Germaine Cellier introduced two partners in crime who'd go on to spawn a whole dynasty of divas: orange blossom and tuberose. Think Naomi Watts and Laura Harring in David Lynch's *Mulholland Drive*: the untapped sexual potency of a fresh-faced ingénue hooked up with the simmering hysteria of an ivory-skinned femme fatale … Cellier also butched up the simpering violet by slapping it with a leather glove (Jolie Madame by Balmain) and invented a whole new perfume family by pouring an overdose of galbanum into Vent Vert (also by Balmain). But she never raised the stakes so much as with her 1944 Bandit.

Conceived in the midst of the German occupation, it is the toughest fragrance ever offered to women; the olfactory equivalent of the street-smart, give-as-good-as-they-get dames of 40s movies. In fact, the blonde, couture-clad, potty-mouthed Cellier could've probably taught a couple of bitchy comebacks to Barbara Stanwyck.

When Fracas nudged me towards her snarling big sister Bandit, I was thrust back to my punk dandy days. Van Cleef had carried some of Bandit's kick-ass genes, and that toughness is what drew me in. But what kept me interested were the film-noirish twists and turns of Bandit's plot: a languid, jasmine, tuberose and gardenia heart, caught up between the earthy green galbanum and bitter artemisia of the top notes and the dark, smoky-leathery base notes – castoreum with its ink and black chocolate facets, the burnt liquorice of isobutyl quinoline, oak moss and smoky vetiver. As though Fracas had slipped on her lover's leather trench coat to slink off on some secret mission. A crawl through a garden, wet earth and grass sticking to her stockings; a tar-roof shed where she shares black-market American cigarettes with a hunted man. Is that a gun in his pocket or …? Bandit may be an outlaw, but she's no femme fatale; in a pinch, she's as good as any man. In a clinch, she'll stub out her cigarette, take that kiss, and growl, 'It's even better when you help.'

Bandit isn't for wusses, which is probably why it was made for women. But if it were launched today, it couldn't possibly end up on the feminine side of the aisle. In the mainstream, masculine and feminine scents are not unlike some species of insects where the male and the female look as though they don't even *belong* to the same species: their evolution has made them diverge so sharply that several classic women's fragrances would smell downright hairy-chested to today's consumers.

Fruit, flowers and vanilla for girls; soap, lavender and wood for boys. So obvious it seems nature-ordained. After all, girls like picking flowers and boys like to whittle sticks, right? Scrap the genetic arguments: men wear floral essences in many cultures

– rose in the Middle East, jasmine in India. And they wore them in the West up to the 18th century: perfumes being mainly custom blends, there was no distinction between masculine and feminine fragrances before Marie-Antoinette's delicately scented head tumbled into a basket.

The great divide yawned open in the early 19th century when upper-class men ditched their coloured and embroidered silks to adopt the black suit as a uniform, leaving it up to their women-folk to showcase the family wealth. Fragrance was contrary to the new bourgeois capitalist ethics, a conspicuous waste of precious materials that evaporated as they were used, as Pliny the Elder was already grumbling all the way back in the Roman Empire. It was also deceitful, suspected of hiding a lack of hygiene or lewd ulterior motives and, as such, clashed with the puritanical values of the Industrial Age. Men were meant to smell clean: the family breadwinner wasn't out to seduce. Though there were, alongside the non-gendered colognes, a few lotions, vinegars, hair cosmet-ics and aftershaves designed for men, the ranges were limited. They did, however, establish what was acceptable for men: laven-der, citrus, aromatic herbs, moss, leather …

The gendering of fragrance was also a consequence of the industrialization of perfumery. Once products started having original names rather than generic ones like 'Eau de Chypre' or 'Eau de Cologne', they had to be geared towards a specific clien-tele. In 1904, Guerlain would put out Mouchoir de Monsieur ('gentleman's handkerchief') to complement its Voilette de Madame, but that was the exception rather than the rule. Men mostly had to make do with what they bought in barbershops, pharmacies or department stores: haute perfumery wasn't meant for them at all.

It was only in the late 19th century that a fragrance family we've come to think of as specifically masculine was introduced, the fougère, (the 'fern') named after Houbigant's Fougère Royale, believed to be the first fragrance in history to use synthetic mate-rials. Fougères are built on a framework of bergamot (an aromatic, peppery citrus), geranium (a rounded note from the rose family, but fresher), oak moss (earthy, green, mossy) and a synthetic called coumarin, initially extracted from the tonka bean (but made much more cheaply through another process), which smells of tobacco, almond and hay. They often incorporate aromatic notes such as lavender and therefore fall within more masculine olfactory codes.

However, the oldest perfume to have been continuously manufactured since its launch in 1889, Aimé Guerlain's Jicky, broke gender boundaries from the outset by oscillating between the fougère and the sensuousness of a fragrance family that was not yet called 'oriental'. Its very name expressed its hermaphrodite nature: it was the nickname of Jacques, Aimé Guerlain's nephew and assistant, but also of an English girl Aimé had proposed marriage to. The girl's parents refused and Aimé never married, or so the story goes.

At first, it is said that Jicky, with its non-figurative name (the first one in the history of perfumery) and ground-breaking combination of lavender, coumarin, geranium and herbs, was considered too revolutionary by Guerlain to be sold to women, though it also boasted floral and sweet balsamic notes, including another new synthetic material, vanillin. When men rejected it, it was offered to women, but it only encountered its public on the eve of World War I, once they had become more accustomed to abstract compositions saturated with synthetics, and just as they were ready to move on from the frills and flounces of the 19th century to Paul Poiret's corsetless Orientalist garb. What Guerlain didn't anticipate was its success with men, who adopted it along with the 1917 Mitsouko, so that, in the 20s, the company had to add the heading 'women's perfumes that could also please men' to its catalogues.

Jicky was, and still is, shared by men and women: any perfume that can boast among its wearers both Sean Connery and Brigitte Bardot or Roger Moore and Jacqueline Kennedy Onassis either suffers from serious gender dysmorphia or sings with the voice of an angel. What's so interesting about it is not only the fact that its notes bridge what we've become accustomed to perceive as the olfactory gender divide, but also that it reconciles the extremes of the clean-dirty spectrum. For centuries, lavender was the olfactory sign of cleanliness: a cultural fact sealed into its very name, which comes from the same Latin root as *lavare*, 'washing' (lavender was used by laundresses, called *lavandières* in French). Whereas its dominant base note, civet (a substance extracted from the anal gland of the civet cat), smells faecal in its undiluted form, in minute amounts it imparts a suave, velvety richness to compositions. Classic French perfumery has always excelled at conjuring the smells fragrances were meant to cover up …

The olfactory codes of the 20th century became settled around World War I. Men were offered scents that conjured either cleanliness (Yardley's Old English Lavender or Mennen's Skin Bracer) or manly activities (the leather-smelling Knize Ten, by the Viennese tailor Knize). The first masculine fragrance that alluded to seduction came out in 1934: the ads for Pour un Homme by Caron depicted either a Greek statue or a gentleman in top hat, white tie and tails. But it still smelled of lavender.

Women, on the other hand, started filching masculine scents in the 20s just as they raided the masculine wardrobe in the wake of Chanel. Caron led the way there too, in 1919 with Tabac Blond, dedicated to the new breed of cigarette-smoking, car-driving *garçonnes*. Chanel followed suit in 1924 by offering them Cuir de Russie. The smell of Russian leather, from the birch tar Cossacks rubbed their boots with to waterproof them, had been known since the 18th century. It was a fashionable way of treating leather, and seems to have first appeared as a note in fine fragrances in the mid-1800s. By the 20s, it had become the scent of the zeitgeist. Paris was teeming with White Russians; Chanel herself was surrounded with them, starting with her perfumer Ernest Beaux. She was friends with the impresario Sergei Diaghilev, the composer Igor Stravinsky, the dancer Serge Lifar; between 1920 and 1923, she had an affair with the Grand Duke

Dimitri. Russia inspired her collections. She borrowed the peasant blouse; she worked with furs; her fabrics were adorned with Slavic motifs by the embroidery house founded by Dimitri's own sister, the Grand Duchess Maria. As for leather, it was traditionally associated with masculine pursuits (cars, aviation, travel), the very activities the 20s *garçonnes* appropriated, and thus became the olfactory emblem of the emancipation of women.

Chanel's Cuir de Russie is actually a variation of N°5. Its warm, slightly oily note is punctured by the fizz of the aldehydes; these thousands of pinpricks infuse the iris into the leather. Propelled by aldehydes, the iris drags all the other floral notes with it: jasmine, rose, ylang-ylang, orange blossom … That is the genius of Ernest Beaux' composition: the subtle feminization of the leather accord, carried out in the same manner as Chanel's hijacking of male sartorial codes.

Throughout the classic era of perfumery, feminine perfumes would continue playing on the range of notes that were considered appropriate for men (leather, wood, citrus, aromatic herbs). Or rather, those notes were considered appropriate for both genders until the early 80s, when powerhouse fougères like Paco Rabanne pour Homme and clear-the-room florals like Poison exacerbated the gender divide, though the pendulum swung back in the 90s with the deliberately unisex CK One or the genderless L'Eau d'Issey. Today, with niche houses refusing to label their scents masculine or feminine and a small number of mainstream masculine launches venturing into floral territories (most notably the remarkable Dior Homme, conceived by Olivier Polge for the then-artistic director of Dior's fashion line for men, Hedi Slimane), the masculine/feminine cursor is a little harder to set. But outside of the aforementioned niche brands, the market is still gendered, if only because the products have to be packaged, advertised and placed on one side or the other of the aisle. The entry-level male fragrances are more apt to be labelled 'body spray' than the wussy Frenchified 'eau de toilette' (cue sniggers as teenage boys joke about 'toilet water'). Since boys marinate in these noxious potions, predominantly redolent of synthetic woods and amber, those are the notes that become encoded as 'male' in the first throes of teenage mating rituals, while girls douse themselves in candied concoctions like Aquolina Pink Sugar or Vera Wang Princess, thus perpetuating olfactory stereotypes that owe nothing to our genetic makeup.

No one is immune to gender categorization and, despite my taste for ambivalent blends, there *are* certain notes that make me feel as though I am wearing male drag. For instance, I can't come to terms with Bertrand's first proposal. The soapy lavender and orange blossom combination draws it towards the fougère family and that's too masculine for me: his part of the story, not mine.

The story of my perfume is both masculine and feminine, of course. A woman in the arms of a man; the fleeting moment when they come together, and the memory of that moment lost. It is also the story of a man and a woman learning to speak each other's language to translate words into scent. *Traduttore, tradittore*, the Italians say: 'translator, traitor'. Is what's going on in that lab a covert war of the sexes?

11

If I know men, I'd say this one was sulking. I've been here for a full five minutes and Bertrand hasn't said a single word since he's sat down to pour alcohol into the mods he's prepared for our sessions – 'mods' being perfume industry jargon for the successive modifications of a scent-in-progress. Perhaps he's a little cross at having to spend time on a non-priority project. He's leaving for Madagascar in forty-eight hours and he's got more urgent tasks, for which he's actually getting paid, by actual clients. Still, when I walked into the L'Artisan Parfumeur boutique, where for some reason he was dealing with a customer, he greeted me with a happy, friendly smile. I guess it may not have been very clever of me to waltz in crowing I had a sample of the next Serge Lutens, especially since every woman in the shop demanded to smell it on the spot. It probably didn't help either that it was a leather scent, like Bertrand's autumn launch, which, being inspired by a journey to Istanbul, also features a prominent leather note. Was I subconsciously exacting my little revenge? When I showed up for our last appointment, he'd forgotten to write it down and we couldn't work, so I just dropped off a Spanish book I wanted him to read.

He hasn't even offered me a seat. I shift my feet as I watch him dip the blotters. The first two mods didn't really speak to me and I'm a bit wary of what's in store. At last, he waves me towards a stool. We jump straight into mods 3, 4 and 5.

'I've toned down the incense and the aldehydes. We're forgetting the soap and lavender, we'll see about them later. I've focused on the floral note. Green. Nectar. Pollen. Narcotic. Indolic. I've played on lily and jasmine. They suit the theme.'

N°3, with its cologne-like effects, is the core of the three new mods: it's an orange blossom in the top, heart and base notes. In other words, they all contain materials that say 'orange blossom' throughout the development. Perfumery materials evaporate at different rates because their molecules are of different sizes: the smallest ones need the least heat to fly off into the air, which is why, for instance, the citrus notes of cologne dissipate very quickly since citrus essences are composed of tiny molecules. The largest, like musk, sit on the skin for hours. So the development of a perfume is like a relay-race: when a material evaporates, another material or another accord (notes that are 'played' at the same time, like a piano chord, and create a different effect combined than individually) takes up the slack to continue the story. The consistency in the development of a fragrance rests on the interplay between top, heart and base notes.

N°4 is more floral, more indolic and spicier than N°3; its spiciness introduces the lily we spoke of during our last session. In N°5, the amber, wood and musk have been amped so that the scent is no longer as much of a floral. But I can't smell the incense. There's a whopping four per cent in the formula, Bertrand informs me, but apparently it's absorbed by the orange blossom because they go well together.

'When you wear it, you'll notice pretty obvious skin and blood effects.'

Blood? How do you even produce a blood note in a perfume?

Bertrand explains that blood's metallic, warm, mineral, salty and rusty facets are rendered through different materials: specific aldehydes, incense, salicylates … The latter generally have a sweet, balsamic, slightly salty 'solar' odour: in Europe, some salicylates were once used as sunscreens until more effective products were found and thus have come to be associated with the smell of sun-heated skin. But Bertrand finds that iso-amyl salicylate also has something matte and metallic about it, like blood. He's kept the costus from the previous mods to produce the effect of veins under the surface of the skin.

'You can't smell veins through the skin, can you?'

'Of course you can. Skin smells the strongest where veins run just under the surface. Here … and here,' he says, touching his pulse and his temple.

'I've never noticed that.'

I'm tempted to go and have a sniff at him to find out.

'Well, when you smell a woman there …'

'… which I haven't …'

'… which you might have, for all I know … Well, in those places she'll smell a bit like warm leather and sheep. Costus conjures that effect. Raw wool. Greasy, dirty hair. It's got a lot of negative connotations in perfumery, but it shouldn't. It's such an amazing product!'

Bertrand rummages in his refrigerator to pull out a phial of costus, does his strip-dipping number and hands me the stuff in a ten per cent solution. I brace myself for a waft of rutting ram but all I'm getting is a rather pleasant, faintly fatty-waxy odour.

'I can't smell much.'

'You can't? You're scaring me!'

'I thought it would be more powerful.'

'It's soft and muffled, but if it's too strongly dosed it can destroy a product, because it gives off a mutton couscous effect.'

'Still, I can barely make it out.'

Bertrand thinks about this for a second.

'You may be anosmic to it because it smells of yourself. You can't smell your own referent.'

How glamorous. I've just been told by a top perfumer that I smell of mutton and dirty hair. Note to self: change shampoo brands. Now let's go back to the blotters, shall we?

I'm not sure about N°3: too bright, too cologne-like. N°4 is turning into a lily. N°5, with its warmer, ambery-musky notes, is pretty sexy and that's the one I'm drawn to the most, but what's it got to do with my story? In fact, what *is* it exactly that made Bertrand see a perfume in that story? I've never asked him in so many words. I do so now.

'Simply the story. I told myself that associating incense with blood and orange blossom would be a challenge.'

But how does he deal with the fact that he's working on another person's experience? When he did olfactory travel sketchbooks like Timbuktu or Dzongkha, he'd actually been to Mali or Bhutan. He knew what they smelled like.

'Stories are pretexts. I start the perfume before leaving. I adapt it and complete it afterwards. The smells you discover on trips can be very striking, but they're usually negative. *Un-ex-PLOIT-able*. You go to a market, you get hit in the face by smells of dried fish, durian, fresh coriander and mango. What do you do with your dried fish? With durian, which smells like baby shit? So you tell the story your way.'

Oh. And there I was thinking of him as an explorer gathering swatches of exotic landscapes to stick them in a bottle.

While we're studying the development of the three mods, I start telling him about my next perfume course in London: I intend to do a comparative study between his interpretation of tuberose and that of another perfumer. As soon as I've said the words, I realize I've just wedged my Louboutin sling-back firmly between my molars. Bertrand scowls.

'It's so *annoying* to see there's a bunch of tuberoses coming out at the same time …'

He's right: the perfume world is barely emerging from a tuberose tsunami as we speak. He shouldn't be surprised, though. If an idea is in the air, more than one perfumer is bound to get it.

'Well, that *pisses me off* ! I'm fed up. What do you have to do to be unusual? *Truly* unusual?'

You can't be *too* unusual, I venture. You need to please at least a few thousand people. So you can't veer too far off the charts.

Soon, we're egging each other on into full-blown angst. We talk about the way luxury giants are killing off the art of perfumery; about the paucity of truly original ideas; Bertrand is even saying, shockingly, that the dream is gone for him.

'What about you?' he asks. 'Are you still *moved* by perfume?'

Actually, I am, every now and then. But I understand how he feels. Sometimes it doesn't seem worth it to add another product to the glut. We speak for a while about the depressive phase artists go through periodically: the moment when it all seems pointless. It's all about breaking down the language we know to find the possibility of another language, I suggest. Like in that Leonard Cohen song that says there's a crack in everything: that's how the light gets in. But you have to go through that destructive phase first. Force the crack open. Bertrand nods.

'It's probably a natural cycle, like plants.'

'And right now you're in the humus, decomposing!'

Well done. Now I've got us both depressed. He needs a holiday and he'll be taking one in a couple of days. I'm not, and I'm eyeing the nearby Pont des Arts with a wistful gaze. Still, the fact that this guy, who's at the top of his game, is willing to disclose his doubts makes him that much more likeable to me. I've been around enough artists to know he's behaving like one right now, even with that sulk about the competition breathing down his neck … time to clear the air about that.

'Anyway, that new Lutens leather doesn't have anything to do with yours.'

'Yes, it does. And that's what pisses me off.'

'At least the beauty editors will have their work cut out for them: they'll be doing a leather theme next autumn.'

'That's just it. Being part of a theme, that's what gets on my nerves.'

I must be doing this on purpose.

'I guess the toughest thing we learn as we grow up is that we're not unique,' I add dismally.

Just kill me now before I go on sabotaging this session.

'Exactly,' replies Bertrand. 'Knowing you're not unique. But at the same time, it's stupid to think that way. It's egocentric. We're not unique. We're here to serve life, not the opposite. Life isn't here to serve the individual. *Never.* You understand?'

What started me on this riff? I'm beginning to suspect that just as he's depressed by the idea that whatever he comes up with, he won't be the only one to come up with it, I must deal with the fact that, to him, my project is one among many, albeit one he took on of his own accord. Whereas for me, it *is* unique: the most exciting thing that's ever happened to me as a perfume lover …

Those three slender strips of paper I've set on his desk are gangways thrown over the sinkhole we've been skirting. It's time we got back to them. I sniff the blotters in turn. The clove-laden lily in N°4 is merrily chewing up the rest of the formula: Bertrand thinks the eugenol – the clove note – slices too much into its volume.

With N°4 out of the running, we compare N°3 and N°5. Bertrand thinks all the different effects – green, mineral, animalic, salty, narcotic – come out better in the former. The base notes, he says, are freer to express themselves in N°3 because N°5 has got twice the amount of musk. As a result, it's the fullest and the most powerful of the three mods, but the sensuous effects of N°3 are crushed: musk wraps the notes beautifully, but the wrong dose can also smother them.

I love the lily in N°5, though: it's a warm, spicy, flesh-eating flower whose dark, honeyed, tobacco-y tones I find compellingly sexy. But my story is about orange blossom. So if the lily is inter-fering, let's chuck it.

'We can play in a more nuanced way with the wood and vanilla,' suggests Bertrand. 'What I'd avoid is the spiciness of the lily. The eugenol almost cancels out the effects of the indole. Both are phenols. Add indole to eugenol and *wham!* Eugenol takes over.'

He'll use N°3 as a base for the next mods, while adding some of the sensuous effects of N°5. Amp up the tobacco note, tone down the green cologne effect …

'The perfume's far from done yet. What's good is that we've prised it apart. We've taken a big step today. I know exactly what to do for the next time.'

Good. Now I can go home and scrub off the mutton grease.

12

D uende. That's what Bertrand and I have christened our perfume-in-progress, though it'll never be called that if it ever comes out: the name has already been taken by the Spanish designer Jesus del Pozo (and to add insult to injury, for an *aquatic* fragrance).

To me, the word epitomizes the heart-rending spell of Seville. I taught it to Bertrand the day I gave him the book where I learned about it, the Andalusian poet Federico García Lorca's *Theory and Play of the Duende*. But of *duende* itself, I'm sure he already knew. For all his boyish sweetness and straightforward-ness, there's a dark streak running through his character, a brood-ing intensity always threatening to boil over as it did during our last session … *El duende* springs from that darkness.

In most of the Spanish-speaking world, the *duende* is a goblin. But in Andalusia, the word is used to speak of Flamenco singers, guitarists or dancers, or of bullfighters whose art moves you to tears. You could translate it as 'soul' or 'the blues'; in a 1999 lecture on the love song, the Australian musician Nick Cave calls it 'the eerie and inexplicable sadness that lives in the heart of certain works of art'. *Duende* is the tragic awareness of death in life and, when artists are possessed by it, their work resonates in ways mere skill or artistry can never achieve. Bertrand's got that, and he manages to express it. It's not by chance incense is one of his favourite materials: his scents are often shot with spirituality or with intimations of death – those earthy, rooty, brackish notes he keeps gravitating to … He's got soul.

While our sessions are suspended first by Bertrand's trip then by his intense work schedule, I return time and again to my samples of Duende 3, 4 and 5. The third, the one we've deemed most promising, is whispering on my wrist in the tentative tones of a thing unborn. It is strange to encounter perfume in this state, terse, barely adorned, unsure of how to introduce itself – in Lorca's words, 'not form but the marrow of form, pure music with a body lean enough to float on air'. The narcotic, honeyed richness of the orange blossom pulled into the mystic, mineral night of the incense; the rusty-metallic flash of blood sullying soft skin.

'All that has dark sounds has *duende*,' writes Lorca, as he explains that an artist can be inspired either by the angel or the muse: 'the angel brings light, the muse form.' But the *duende* is more than just a source of inspiration: it is a demon the artist struggles with, 'not a question of skill, but of a style that's truly alive'.

Should a perfumer surrender to his *duende*? Probably not. Perfume is meant to charm like the muse or dazzle like the angel, not 'burn the blood like powdered glass'. But perfume can have *duende* since, like music, dance and spoken poetry, 'the living flesh is needed to interpret [it]' with 'forms that are born and die, perpetually'. Is perfume not a perpetually moving space of beauty borne by living flesh?

Duende is the blade that tears beauty away from you even as you experience it most fully; it is the becoming-a-memory embedded in each moment of grace; a thing that is real and yet an ungraspable wisp, just as perfume conjures both the presence and absence of a loved one, a moment of the past, a persona worn and shed. An object that consumes itself as you experience it, whose very enjoyment signifies its annihilation, except as an ever-shifting imprint on the soul …

So now is the time to reach into that hallway closet and to pull out what's slept there for two decades. A Venetian notebook with a pattern of cream and green fern leaves. And a black glass bottle with a frieze of nymphs.

The first was given to me by a French writer. I'd go up to his flat on Sunday afternoons, hurriedly removing my knickers in the lift because the bedroom always came before the whisky and the talk and the laughs in his study. One day, handing me the notebook, he'd said: 'Write everything that happens to you and more will happen,' as though the very fact of writing would generate more adventures, and it had, and it has.

The second, I found within days of arriving in Paris, where I had come to live at last. Officially, it was only while I did my PhD on 18th-century French literature. But I knew that Paris would be my home; that I would define myself by my destination rather than my origins. I'd stumbled on this new perfume in a knick-knack shop called Divine near Saint-Sulpice, one of the very few baroque churches in Paris, where both the Marquis de Sade and Baudelaire were christened. It became the scent of anywhere-but-Montreal, the scent of leaving home, the scent I wore when I walked into beauty …

As I open the notebook I filled out for two years and have never touched since, I breathe in the perfume I wore for much longer than that before rejecting it utterly, to the point where – shades of my father – smelling it made me queasy, though I'd kept the bottle for memory's sake. First a squirt on a blotter. Then a tiny spritz on my wrist. And then, more generously, on my breastbone. Dusky, powdery rich tobacco-hued vanilla, run through with austere, rooty vetiver … No flashback: this was part of me for too long, throughout too many adventures, to be associated with a particular period, mood or event. What I get is … comfort. Like meeting an old friend and resuming a conversation interrupted years before. I'm surprised to catch myself thinking I could wear this again. But somehow, not surprised at all to find out that on the very first page of the notebook given to me by my lover-mentor, one of the very first words I wrote was this one: Habanita.

'Habanita. Hasn't my life, since I've encountered it, been wholly turned towards this perfume? I feel as though I'd spent my life going towards it, in the opposite movement of those women who distance themselves gradually from the perfume they've chosen for themselves, or of those who hang on to theirs because they're afraid of change.

Habanita. Little by little, I have accomplished this perfume.

I smoke Havana cigars now. I wear carnations in my hair. I wrap myself in dancing silk dresses that cling to my breasts and my hips. Since Andalusia, I want to feel myself thus, heavy, hot and a little damp in a dress that caresses my body.

Habanita: Spain. Seville. The place where I found my perfume. But also incense and orange blossom.'

I wrote this over twenty years ago after coming back from my second trip to Seville. It was all Immanuel Kant's fault, really. I'd read his *Critique of Aesthetic Judgement* in one sitting for a philosophy seminar and been struck down with such a violent fever I became delirious – my writer-lover said that was what happened when women like me were exposed to German philosophy. My fever-dreams were so haunted by images of Seville that as soon as I managed to crawl out of bed I booked a trip there for my Easter holiday: the Catholic-pagan rites of Holy Week were the only possible antidote to the philosopher whose habits were so regular, the legend goes, that you could set clocks by his daily walks.

I'd first gone to Seville the preceding autumn with my parents, just before settling in Paris. Our car had broken down right in front of the hotel so my first day there had been spent in a garage, acting as my father's interpreter with my high-school Spanish, watching a young mechanic, slender wrists emerging from the sleeves of his green overalls, lazily tugging at charred, greasy things under the hood. Spying the way he stared at me from under his fringe, as frankly and guilelessly as a dog. What did he see? He couldn't possibly read all the signs I'd accumulated on my body, the signs behind which I retreated, the signs that said 'intellectual', 'sophisticated', 'scornful of the clichés of femininity'. By then I'd ditched my New Wave garb to go post-apocalyptic Japanese – asymmetrical haircut, body drowned in geometric sack-dresses, feet shod in flat clunky shoes. But this boy was looking at me as though I were a woman. No signs, no words. Just a woman.

First I cringed. Then I craved it. And somehow, the disconcerting pleasure of being just-a-woman rather than a Parisian intellectual, a creature capable of decoding Jacques Derrida before lunch and still have appetite left over for a nibble of Jacques Lacan, carried over into my new Parisian life after my Spanish holiday. As I flitted from seminar to seminar, I started studying another, parallel curriculum. I found Habanita. And I followed its sensuous wafts all the way back to Seville.

So I returned the following spring to see the celebrations of Holy Week with my writer-lover's cream and green notebook tucked into my bag, the phone number of a friend of a friend, the address of the elderly aristocrat who was renting me a room and a bag full of the vintage frocks that had been steadily gaining on deconstructed Japanese slipcovers in my closet at the Cité Universitaire. My lone phone number yielded a whole tribe of opera-loving, aristocratic aficionados. I was young, I was alone, I knew all the operas they talked about, from *Don Giovanni* to *Carmen*, and I could sing a few arias in smoky bars: they adopted me as their new mascot, a nubile stranger dropped into their midst to churn up a little erotic excitement. Alberto suggested I paint a beauty mark under my left eye; Jacobo planted carnations in my hair; Perico gave me my first Havana cigar to light up when the second bull came out and smoke until the sixth was dead; Pepa taught me to handle a fan, sitting on the grass under the Puente de Triana.

And Román picked me up as I was crossing it, a tall slender-hipped young man in a black suit with a green polka-dotted handkerchief tied around his neck, gypsy-style. Román who gave me the most beautiful night of Holy Week, the incense and orange blossom and beeswax candles … At the time, my elderly hostess had warned me he might be a pimp, but does a pimp tell you 'I spoke to you because you had a delicious little 18th-century air about you'?

Paris had taught me about perfume. But it was in Seville that I learned about scent in an intimate, carnal way, experiencing with my full body rather than from bottles. The whole city was possessed by smells, its air thick and churning with them. The mere act of breathing was a constant intoxication.

Orange blossom exploding into a scented atomic mushroom in every street and square. The lilies that spilled from the altars of the churches where I sought out coolness after wandering in the streets; the pungent, sweet incense soaked up by the golden, swirling baroque altars, blending with the beeswax of the pews and candles. The cold, musty gusts exhaled by old palaces; the moist, vegetal smells seeping through the grilled gates of fountain-cooled patios … Jasmine nebulae spilling over from walled gardens, so narcotic they made my knees buckle, so beautiful they drew tears. In the evening, the gypsies would go from bar to bar, baskets filled with bunches of jasmine blossoms stuck on\a pin, and the men would buy them to wear on their lapel, and you'd tuck one in your curls, and the next morning even your pillow would be fragrant. And the tall, thick-stemmed sprig of trumpet-like blossoms that Teresa brought to me one day, the one she called *nardo*, as heady as jasmine but with a sharper, cooler smell so strong it wafted out from my tiny hotel room onto the roof patio where the gypsy chambermaid came to hang the wash and drew her to my window: my first-ever encounter with tuberose …

The food too was fragrant. The bitter leathery saffron of paellas, the sweet fatty batter for fish, the honeyed smoky ham made from pigs fed on acorns, the pungent mutton broth spiked with fresh mint, the sticky cakes soaked in orange blossom water and sprinkled with cinnamon inherited from the Arab tradition. And the manzanilla wine that came from Sanlucar de Barrameda where the Guadalquivir threw itself into the Atlantic ocean, which carried in each drop the memory of the saline winds blowing from the marshes on the other bank … And the very streets, heated asphalt rubbed slick with the drippings of thousands of beeswax candles which could never entirely be blasted off by the municipal workers with their hoses, melting in the heat to make the streets slippery and honeyed. The stale-water stench rising from the arm of the Guadalquivir which ran through Seville – at the time, it was cut off to stem floods, and it festered in the heat. And the hay-spiked dung in the trail of the horse-drawn carriages that stood waiting for tourists around the cathedral, warm and homely. And the carnations gypsy fortune-tellers would hold out to you before grabbing your hand to predict love with a dark-eyed stranger. How could that prediction not come true? The city was teeming with dark-eyed men and I hadn't spent three days there before one of them found his way to me …

It was in Seville, as I wandered the streets from church to café, that I started weaving the threads that bound me to Habanita.

Habanita: Little Havana. Havana: cigars. Cigars: Carmen the *cigarera*. The very archetype of the fiery Spanish woman, except that she isn't Spanish at all since she was invented by a Frenchman, the writer Prosper Mérimée, and turned into a myth by another, the composer Georges Bizet. Sevillian girls called Maria del Carmen often added, when introducing themselves to foreigners: 'But I am Carmen, *la de España*, she of Spain, not *la de*

Bizet.' Fake Andalusian that I was with my cigar, fan and carnations, I was frequently photographed by tourists, perhaps the most authentic Carmen in town. I knew how to sing the 'Habanera' and the 'Seguidilla' – I'd never known until then I had a good singing voice, but I did, and I used it. And I was nicknamed Carmen, *la del Canada*.

And then there were the 20s, the decade I was most drawn to, not least because Hemingway's *The Sun Also Rises* was set in the 20s. I identified with his Lady Brett Ashley, the expatriate, bob-haired, hard-drinking femme fatale who'd gone off with a bull-fighter in Spain. And because the man who gave me my nickname 'Carmen, *la del Canada*' was an old bullfighter who'd actually been a friend of Don Ernesto's: in fact, the very son of the bull-fighter depicted in *The Sun Also Rises*. It couldn't be a coincidence. Seville was dotted with buildings from the 20s, left over from the 1929 Ibero-American fair. That junction between the Old and the New World was one I was living in the flesh. So choosing a perfume from the 20s, one that was named after a former Spanish colony, couldn't be a coincidence either.

Habanita was launched in 1921 by Molinard, first as a fragrance added to cigarettes (Carmen again) – women had just taken up smoking but they didn't want to reek of stale smoke – and that too, that sign of feminine emancipation, became part of my story as I dived alone into an unknown city. I would've told myself any story to turn my attraction for Seville, Carmen and the *années folles* – the 'wild years', as the French call the Jazz Age – into some sort of destiny. I blamed Habanita. 'The most tenacious perfume in the world', as it was advertised in 1924, had wound its way into my blood and transsubstantiated my very flesh. Habanita was five years of my life, years that found me drawn time and again to Seville, where I would spend months on end, waiting for adventures to happen – and often they did.

* * *

As I breathe it in from my wrist, this scent I had pushed back to the furthest recesses of my memory just as I had tucked the bottle in my darkest closet, I suddenly realize what made it possible for me to re-open that door; to love a type of scent I'd rejected along with Habanita when I had to give up *being* Habanita. That small atomizer has been sitting on my dresser all along. It was the first scent Bertrand and I discussed at length, the one he used to explain to me how he worked. The first vanilla I could wear after two decades of vanilla avoidance conjured my long-lost Habanita through its notes and original name (it has since been renamed Vanille Absolument). Who knows whether it wasn't that subconscious connection that inspired me to tell Bertrand about Seville? Though I'm not quite sure I believe in fate, I *do* believe that if you're alert to signs, able to connect them and willing to follow their call, strange patterns emerge that do look a lot like fate. It's called a story. And in this story, I'm letting my nose lead me: in French, *flair* means both 'sense of smell' and 'intuition'.

Is the ghost of Habanita, with its orange blossom, amber, vanilla and musk, the reason why I am still so drawn to the dark, velvety Duende N°5? Though we ultimately rejected it because the lily veers too far off my story, I keep going back to it. But despite their common notes and a similar sensuousness in their textures, Habanita and Duende N°5 clearly come from different universes: the latter doesn't have a retro molecule in its body whereas the former is powdery and spilling out of its décolletage as only perfumes composed before the 60s could be.

Duende N°5 is only a sketch, of course, which explains in part why it feels so much more modern. Yet as the orange blossom top note gives way, it also feels much more *ancient* than Habanita. Perfumes in the 20s *did* look back to the faraway origins of the art, to the balsams and resins of Oriental perfumery, as though the sleek lines of the flapper's shift dresses, their cropped bobs and naked limbs, called for more elaborate ornamentation to compensate – hence the stylized embroideries, heavy makeup and heady, come-hither fragrances of the period. But they didn't take that final step into the most archaic of all scents … Incense.

13

Incense is one of the oldest continuously used aromatic materials; perhaps the first letter of the perfume alphabet. For centuries, it has been the language men speak to the gods as well as the physical experience binding together the faithful; the invisible realm made perceptible to mortal senses.

The word, derived from the Latin *incendere*, 'to burn', is practically synonymous with perfume, which comes from *fumare*, 'to smoke'. Most ancient cultures burned incense: Egyptians, Assyrians, Babylonians, Hebrews, Greeks, Hindus, Chinese. Even today, from Saint Peter's in Rome to the Buddhist temples of Kyoto, and from the *diwans* of Oman to the temples of Benares in India all the way to Californian meditation centres, incense is still burned all over the world. The practice is both so archaic and so alive that the smell stirs ancestral memories even in those who aren't drawn to spiritual pursuits.

In various cultures, the word equivalent to 'incense' can designate aromatic resins, woods from various trees or complex blends, from the dazzlingly refined compositions of the Japanese art of Kōdō to the sickly joss sticks burned in Chinese restaurants. But the incense we're putting into Duende is the real stuff:

the resin of the tree *Boswellia carterii*, known as olibanum or frankincense in the Bible.

Bertrand has told me that, to him, incense is blood, and many traditions bear out this symbolic association. Incense *did* mingle with the blood of ritually slain animals or human beings on countless altars. In fact, the Ancient Greek word for sacrifice, *thuos*, originally meant 'substance burned to obtain fragrant smoke': it encompassed both the aroma rising from the roasted carcasses of sacrificed animals (the Olympians fed on the smell while humans feasted on the flesh) and the scent of incense. But incense was also literally thought of as the blood of the trees that gave away their fragrant resins through wounds in their bark, as the anthropologist Annick Le Guérer explains in *Le Parfum: des origines à nos jours*. Those trees themselves were often believed to be supernatural beings, humans or nymphs metamorphosed through divine intervention, so that, again, spilling their resin was literally spilling blood. The very beauty of their fragrance was proof of their holy nature. Did the gods not exhale sweet scents? Perfume was the vital principle of blood in its purest form; the very essence of sublimation.

We've come down from Bertrand's lab, where he's spilled a phial of a particularly diffusive material that would cover up pretty much any other smell, to discuss Duende N°6 and 7 on the terrace of a nearby café. The roar of the traffic rushing on the Quai du Louvre is deafening; gusts of dry wind blow the minute blond hairs of plane tree flowers into our eyes and noses. Bertrand complains he has trouble smelling anything in this weather. Imagine what it's like for my untrained nose. So we've been veering off on tangents. I've wanted to question him for some time about his relationship to incense. The first perfume of his I wore, back when I hadn't the least idea who he was (nor any perfumer, for that matter) was put out by the Japanese brand Comme des Garçons in 2002. Avignon, named after the city where the papal court took refuge in the 14th century to escape the clan conflicts that raged in Rome, was such a striking evocation of Catholic incense that I couldn't wear it without feeling faintly sacrilegious.

It's not quite clear how far back the use of incense goes in Christian rituals. Some fathers of the Church condemned it because it was widely used by pagans: persecuted Christians were forced to burn a few grains of incense on the altars of Roman Emperors who were worshipped as gods – it was either that, or off to the lions. But it is mentioned in the Gospels and by the 5th century it was officially part of Christian liturgy. It was certainly a welcome cover-up for the miasma of the unwashed masses, mingling with those rising from graves both inside and around churches. But it had a deeper, mystical meaning: it symbolized the prayers rising to God and the sweet scent of Jesus' words penetrating to the very souls of the faithful. The blood of the olibanum tree was a metaphor for the blood spilled by Christ. Catholic liturgy thus acknowledged and appropriated a symbolism that went back millennia …

> Mayest thou be blessed by Him in whose honour thou art to be burned. The tears of a wounded tree are twice blessed in the Mass. Twice blessed therefore is the creature of nature which, being wounded, gives up its fragrant tears in honour of Him who wept over Jerusalem; in honour of Him who was wounded and shed His precious blood for the whole world; in honour of Him whose unbounded love extends to all nature. All nature in turn serves Him, but the tears of olibanum are twice blessed.

The friend I was with when I bought Avignon, like me a pure product of Quebec's Catholic educational system, was impressed by the accuracy of the rendition, though she tried to dissuade me from buying it. 'Why would a woman want to smell like a *church*?' (*That* was rich from a woman who'd once bought a cassock so she could dress her lover up in it …). But I sniffed Avignon obsessively. It carried me back to Holy Week in Seville, my first full-body immersion in church incense, since when I was a child Masses in Quebec involved guitar-playing priests in Frank Lloyd Wright-inspired buildings rather than the imposing liturgy of the Old World. I used to burn actual church incense on charcoal tablets, which I bought in a Catholic geegaw shop on the place Saint-Sulpice. I stopped when a neighbour, alarmed by the smoke seeping out from under my door onto the landing, called the fire brigade. Delighted as I was to find a couple of brawny French *pompiers* in their skin-tight black uniforms getting ready to break down my door, I forewent church incense from that day on.

Avignon was one of the fragrances that put Bertrand on the map and it enjoys a cult status. Since then, incense has been a leitmotif in his work, and pops up in at least one composition out of four, which amounts to something like an obsession.

'You're right,' he nods. 'I've always been fascinated by it. It could be my Catholic upbringing coming out, my childhood spent in churches. The smell of incense blending with the smell of old, damp stones ... It's the only thing I've kept from Catholicism. The rest, I've rejected. It brings back too many bad memories.'

As Bertrand reminisces about the strict 'Old France' education inflicted on him and his siblings – his grandmother, he says, was as handy with the horsewhip as with the Bible – I suggest his vocation as a perfumer may have been a way for him to break free from it. By re-appropriating incense as a sensuous pleasure, he's managed to exorcize painful memories. After all he could have rejected it just as passionately as he says he's rejected his religious upbringing, though there is something religious about the term 'vocation' ...

'Exactly! It's a priesthood!' he chuckles.

I tell him about my own phase of incense madness. I chewed the small resin chunks (they tasted soapy) because I'd read somewhere it would make me exhale the scent through my pores. I stopped because I was afraid that the resin would clog up my gut and land me in A&E. Intestinal obstruction would have been a far cry from wafting all the perfumes of Arabia. Though the punchline makes him laugh, he cocks his head, dead serious:

'So you have a very mystical side to you.'

'Do you think so?'

'I'm asking you.'

He's really in earnest. In fact, he's one of the most earnest people I've ever met, almost like a child. His question sets us off on one of those involved, intense discussions that always seem to develop when we're together. Hunched over our blotters, we go on about the way poetry and artistic creation fatally lead us to wonder about transcendence. We're both cold sober and it's 6 p.m., but you'd swear this was one of those rambling philosophical talks you fall into at 6 a.m., after you've dug up that bottle no one ever bothered to open, the one with the weird label on it, because you were all out of whisky. I end up pulling out a 1918 prose poem by Pierre Reverdy, 'The Image', which I had printed out. In it, the poet claims that 'The more remote and accurate the connection between two realities that are brought together, the stronger the image – the stronger its emotional potential and its poetic reality.'

I feel that's what he's done with my story, I tell Bertrand: listening to it, he connected orange blossom and incense in ways both 'remote and accurate'. I'd registered it like a sensitive plate. Twenty years later, he saw it intuitively. His senses had processed the connection even before he became aware of it; then he figured out rationally that their common mineral notes were what drew orange blossom and incense together.

'You're right. It's a gut feeling. I don't always know where I get that stuff. I guess the body knows things you don't.'

'But those moments of grace can only happen when you master your technique perfectly. Like a dancer …'

It's still all about the *duende*, isn't it? 'Not a question of skill, but of a style that's truly alive', as Lorca writes. We both pensively duck our noses towards the blotters. Despite the raw wind and exhaust fumes, what's rising from N°6, creamy and petal-fleshy, is drawing me in. Bertrand explains he started out with N°3, which was the most obviously orange blossom/incense, and tried to keep that duality while making it 'more attractive and modern still, more floral and headier', adding a heady, fruity, full-flower jasmine, 'so delicious it's almost like *banana*'.

N°6 is quite green. I thought we were toning down the green? Bertrand says he took out the green note that smells of hyacinth and ivy – it's not our story, not a Spanish spring – but not the one that smells like sap and pollen and feels moist. He's also diminished the percentage of incense. In N°7, on the other hand, he's kept a higher dose of incense but counteracted its harsh effects by adding even more green and boosting the fruity facets of the jasmine: 'almost strawberry jam, banana jam', he says. He's also added beeswax and honeyed, pollen effects, 'to bring out the fleshy side of the flowers', and worked on an amber tobacco base, reinforced by immortelle, 'to bring out the cigar effect'. Immortelle comes from Spain, he adds, so it's part of the story.

But I'm a little disappointed not to find the velvety, woody accord that I kept coming back to in Duende N°5, I tell him.

'I dropped it completely. Zero. You told me to go for N°3. I can work it back in if you want, but not as strong, because even though it's vibrant it skews the story too much. You lose the orange blossom note. Maybe I'll amp the incense back up … I did these yesterday, I don't have any distance. We've got to see what effect they have on skin.'

I suddenly remember I was intending to ask him about the blood note. I don't know how it came in. As far as I recall, I never mentioned it in my story.

'But of course you did!' he bristles. 'We were sitting right here, at this very table, when you talked about it, the day you gave me the book by Lorca!'

That was the day he'd forgotten we had an appointment and we just went down for coffee. I must have blocked out our conversation because I was a bit upset, but now that he mentions it, I do remember reading out parts of the book to him. That's when the name of the perfume went from 'Séville Semaine Sainte' to 'Duende'. I've noticed that Bertrand always gives evocative names to his scents-in-progress: for instance, the Nuit de Tubéreuse he gave me a sample of the first day I came to see him started out as 'Belle de Nuit'. The name *is* the idea, he tells me, and the idea must be absolutely clear from the outset. The orange blossom-incense-blood accord sprang from the notion of *duende* as soon as he'd read Lorca. Since Holy Week *is* about the Passion of Christ, there was already blood in the story, so it makes sense.

'We could add the smell of ashes,' he muses.

'That would be going back to Ash Wednesday, and the story unfolds during the night from Maundy Thursday to Good Friday ... But I did tell you about the smell of old stones that came from the church ...'

'Ah, yes, we could work that in ... But no, we'd be falling back into stuff I've already done so many times. I'd be repeating myself.'

'Then let's not. We want something new! I get the feeling this isn't like anything you've ever done before, is it?'

'Oh no, it isn't. Absolutely not.'

14

'Do you know what you want?'

Today's our last session before the traditional French full-month August holidays, and I'm seething. I've lived in Paris half my life but I've never become used to the whole country coming to a standstill for a month. Like half his compatriots, Bertrand will soon be waltzing off to the South of France and he's in a rush to wrap up – again – so he lunges into technical explanations without any foreplay.

The three mods he's just mixed, numbered 8, 9 and 10, are tweaks on Duende N°6: he's added different green floral notes in varying doses. Algix, Canthoxal, Lyral, Lilial … As he bombards me with the names of molecules, I realize I'm tuning out. I'd need to smell them to understand what he's saying. So I wait until he's done with the chemistry lesson to ask him the question that's been needling me lately: what's the current state of the perfume compared to a finished product?

'It's subjective,' he answers. 'We might be practically done. Or consider that this is just the starting point. Some perfumes were great successes based on one or two mods. Others were monumental flops after going through five hundred mods over two years, with teams of perfumers working on them. The main thing is to know when to stop. To know how to choose. You need to know exactly what you want.'

And that's when he pops the Question.

'Do you know what you want?'

Indeed. What *do* I want? Do I have any idea of what Duende needs to be? Maybe I've just been going along for the ride, commenting, storytelling, philosophizing; entwining threads of words around his work … That's what I always do: find a door that's open, walk through, enter another person's world, try to make sense of it, to capture it in words. Writers are greedy that way. Usually, what I ask from a perfume is to be taken elsewhere. This time, I'm the one who's taking the perfumer to a place he's never been. Bertrand, are you following me? I'm lost too …

After a few seconds of silence, Bertrand bursts out into a teasing laugh. Affectionate, but only just.

'*I* know very well where I'm going. I'm starting to get the accord that will be about ninety per cent of my formula. I'll embroider on the remaining ten per cent. Work on the top and base notes.'

I'm still mulling over his question. He's thrust me into a position that I have neither claimed nor sought out.

'Hey, you know what? I'm not your client. I'm not a brand owner, I'm not a project manager, I'm not an evaluator and I'm not commissioning a bespoke perfume. That's not my position. I've handed my story over to you and now I'm registering how it develops.'

'Still, you need to be able to tell me whether the perfume is adapted to your story.'

He's not about to let me off the hook. Have I been failing him somehow? I haven't even told him how beautiful I found what he was doing, especially the last floral accord we agreed on, N°6, which I've worn frequently over the week.

'I think we've started getting there since the last time.'

'Since I came up with N°6. Because the accord is already becoming good. That's fundamental. We've got to create something that stands up. That's original but pleasing. We've got to avoid segmentation. We can't afford it. Otherwise, we'll please a small number of aficionados but our perfume will never *live*.'

A perfume needs to fulfil three fundamental criteria so that it lives, and lasts, and goes on gaining recognition as time goes by, he adds.

'One: originality. Two: diffusive power. Three: tenacity.'

This is the first time I've heard Bertrand use a marketing term like 'segmentation'. The first time he mentions commercial success. Of course he wants his perfumes to sell well. This is how he makes a living. If ours never goes into production, he'll have invested as much time and effort as for the stuff that *does* make it to the shelves. Even if he recycles his ideas, it's still a frustrating process. I can't blame him for wanting this one to be successful, should a company decide to commercialize it. But there's more than profit involved. He wants his stuff to be loved. To be worn. To endure. Perfumers know their work is heartbreakingly ephemeral. Many products don't even make it past their first year. And since there's a bit of my soul in that bottle, I too want Duende to survive …

Meanwhile, it's not Duende's chances of making it into the 22nd century that concern us, but how long it lasts on skin. If I'm not up to Bertrand's standards as a project manager, at least I can serve my purpose as a human blotter. Duende N°6 doesn't have enough staying power, so it isn't fulfilling the third prerequisite, tenacity. I've had compliments on it, so the diffusive power seems to be satisfactory. But a friend of mine told me: 'If this perfume is meant to be you, it needs to be darker, more sensuous.' Bertrand nods.

'Interesting. So we'll play up the sensuousness of the base notes, to give it more mystery and more persistence. This will inflect the global accord by five to ten per cent.'

'I still think there's something to pull out of that sensuous base I loved in N°5.'

'We'll go back to it if we need to.'

'Because it reminds me of Habanita, somehow.'

'What does?'

'That wood, tobacco, musk and vanilla base reminds me of Habanita.'

'You want me to play on that? Is it related to the story?'

Haven't I told him about ten times?

'It's the perfume I wore back then.'

'So it's important.'

'I've got an old bottle which probably goes back to the mid-80s. There are just a few drops left at the bottom, it's like liqueur.'

Bertrand nods vigorously, eyes twinkling.

'Brilliant! Bring me Habanita! I *want* it!'

One week later, after smelling the new mods at leisure, I fire off a text message to Bertrand, who must be somewhere in the area of Grasse: 'I know what I want. Don't work on anything before we see each other again.'

But what I really want to say is this: Bertrand, this isn't my story. Not yet. The orange tree is there. But gorgeous as it is, there's no one standing under it. My body pressing against Román's, the scent of tobacco on his skin, his hand under my skirt ... Duende should make me want to say a prayer and get my knickers ripped off, *at the same time.*

15

So now's the time to bring back Habanita and all she stood for; that thing I lost, through the sheer exhaustion of holding myself ready for adventure, of being Carmen.

I don't remember the exact moment when Habanita started triggering migraines, nausea, anxiety. But it happened. There was a man, of course. The Tomcat turned up as strays always do when they sense there's a home that's open to them. With his messy shock of hair, long nose and craggy face, he was sexy in a mountaineer sort of a way. Nearly twenty years older than me, but didn't look it: I suspected his enduring youthfulness came from the fact that he'd never held down a suit-and-tie job, or any kind of a job for very long, really. He'd been a ski monitor, a reporter, a stage manager for avant-garde directors. He'd sold vintage clothing and published a novel; when I met him he was writing screenplays for television.

By that time I'd drifted away from my PhD into freelance journalism, writing profiles, travel pieces and columns for a trendy French left-wing monthly. I'd never really seen myself as an academic. Besides, there were just too many parties in the 80s, too many private views, too many movie houses, too many trains to take me across Europe – and journalism got me where I wanted to be, backstage, flitting from subject to subject rather than focusing on a single field. But it didn't pay well so, when I got too broke to make the rent, my Siamese and I moved into the Tomcat's small flat overlooking a leafy cobblestoned court-yard. When I cried because my best friend was pregnant, he said we could have a baby, and that he'd marry me. So I unlearned the way to Seville and stashed away my black bottle of Habanita.

Bereft of my signature fragrance, I set out to find one that would reflect my new identity. In Habanita's arms, I'd managed to sail past the power-suited 80s; by 1990, the likes of Poison and Giorgio Beverly Hills were starting to feel as trite as the shoulder pads I was ripping out of my jackets. Besides, one of the reasons I'd picked Habanita to start with was that it was relatively obscure. It felt like my secret, and meeting another woman who wore it was like finding a long-lost sister. So I decided to explore lesser-known fragrances from the same era. Reading a biography of Anaïs Nin, I found out that she'd worn a perfume called Narcisse Noir by Caron, a name already famil-iar to me thanks to my childhood idol Geneviève, who spoke admiringly of its best-selling Fleurs de Rocaille. I'd start with that.

The Caron boutique on the corner of the avenue Montaigne and the rue François 1er, a short bus ride away from my new home, was a jewel box full of rose- and violet-smelling face powders, pastel-tinted swansdown powder puffs and giant 18th-century-style crystal urns with golden faucets from which perfumes could be poured into rectangular bottles of various sizes.

Like Chanel's Ernest Beaux, the founder of Caron, Ernest Daltroff, was Russian-born; like François Coty, the Napoleon of Perfume, he was self-taught. But while Coty decorated his chateaux in the style of the 18th century with a social upstart's tastes in art that contrasted with the modernity of his products, Daltroff, who came from a wealthy family, was keenly interested in the contemporary art scene. The vivid, unnatural colours and violent contrasts of the Fauvists, as well as the olfactory memories of his exotic journeys, inspired a similarly vivid, unnatural palette, explains the perfumer Guy Robert in Michael Edwards' *Perfume Legends*. Daltroff was unafraid to experiment with the new, powerful materials that more seasoned perfumers shied away from. The emotional intensity of his compositions, their sheer *gaudiness*, captured the very essence of an era when the avant-gardes, from Diaghilev's outrageously erotic and exotic Ballets Russes to Picasso's grinding up of traditional perspective, shattered the remnants of the 19th century. The house of Caron carried its own brand of Russian revolution into perfumery.

It was the first time I'd set foot in a historic perfume house and discovered a series of fragrances composed by the same person; though my nose was still uneducated, I could sense a common thread: a dark powdery mossiness which I would learn later on was called the 'Mousse de Saxe', a base created nearly a century ago in Grasse by one of the unacknowledged geniuses of perfumery, Marie-Thérèse de Laire, who worked in the family company. Bases were mini-perfumes conceived to dress up new molecules that could be perceived as too harsh or unwieldy, and were incorporated directly into perfume formulas. The Mousse de Saxe had been elaborated around a mossy, leathery, liquorice-like material called isobutyl-quinolin. It was at once sweet and bitter, powdery and green, rosy and spicy, as bold as the vibrantly coloured Orientalist styles the couturier Paul Poiret made fashionable before World War I ... I've since learned that the Mousse de Saxe was an inspiration for Habanita, but even without knowing it, I knew I'd found a fragrant home. The Caron urns were time machines carrying me off to the early decades of the 20th century, each drop of fragrance conjuring visions of a glamorous past. N'Aimez que moi ('Love only me') was given to their sweethearts by men departing for the trenches during World War I; En Avion was a tribute to the chic aviatrixes of the 20s; French Cancan was meant to draw in the

American clientele in the interwar years with a name evocative of Gay Paree.

I decided to stick to the scent that had prompted my visit and departed in a cloud of orange blossom laced with dark, dirty notes that conjured silk underpinnings shed during a particularly decadent party in some silent movie star's Hollywood villa. The Tomcat took one sniff and vetoed Anaïs Nin's Narcisse Noir – I smelled like the most popular brunette in a Pigalle brothel, he snorted – just as he'd rejected the lingerie collection I'd assembled over the years. He'd also vetoed my portrait by the photographer Bettina Rheims, taken to illustrate one of my articles: dressed solely in a black lace Merry Widow, I cupped my breasts, head thrown back, lips parted … Bettina had deemed it good enough to be part of her *Female Trouble* book and exhibition and to hang a print of it in her flat. But the Tomcat loathed it. So Bettina joined Habanita and the Merry Widow in the closet, and Narcisse Noir was forgotten. I guess I was grateful to the Tomcat for seeing past the man-eater image I'd been peddling before meeting him. Men don't marry the Carmens of this world. They kill them.

My next Caron sample, however, met with his approval. Farnesiana was named after Michelangelo's Farnese Palazzo, the French embassy in Rome (and because perfume can be a form of prophecy, there would be a July 14th party at the Farnese one day, though not with the Tomcat), but also after the *Acacia Farnesiana*, a relative of the mimosa, with similar tiny yellow powder-puff flowers. It reminded the Tomcat of his summers camping on the Côte d'Azur with his parents in the early 60s: that was when he'd dated an English girl who'd gone on to become an erotic icon of European movies in the 70s, a memory he treasured. Farnesiana's powdery, almost edible tenderness – it smelled of marzipan, violets and, more faintly, of anise – was like Habanita's sunny side. It also reminded me, oddly but endearingly, of wet Kraft cardboard and of the Colle Cléopâtre whose almond aroma was so tempting generations of schoolchildren snuck a lick of it when the art teacher wasn't looking.

I'd found the scent that would take me to my wedding day, a good-natured affair in the city hall of our Parisian district: the Tomcat, a libertarian child of the 60s, had flatly refused a church ceremony. Until, that is, the Caron boutique manager slipped a sample of Poivre in my bag. I dabbed it on. The Tomcat grumbled it smelled like a dentist's surgery. But something had stirred in the pit of my stomach. Though they exuded a similar powdery retro charm, Poivre was the polar opposite of Farnesiana. It didn't smell of its eponymous pepper, but of cloves and the spicy red carnations I'd so loved in Seville. If Farnesiana cooed reminiscences of springs on the Riviera in dainty, Grace Kelly-style New Look frocks, Poivre, like Ava Gardner in a foul mood, could slap you just for kicks. In fact, its eau de toilette version was called Coup de Fouet, 'Lash of the Whip'. In French, something that gives you a *coup de fouet* revives your energy. But the kinky subtext suddenly brought back memories of a lover with whom I'd indulged in an experiment that had left me unable to sit – but gloating at my own naughtiness – for a couple of days … So I came home with Farnesiana's hellcat of a sister. And that was just for starters. Coup de Fouet soon loomed over my two other Carons. I ignored the Tomcat's grumblings: by that time, I'd nabbed a full-time job on a women's magazine and spent most of my time at the office anyway.

Then when Narcisse Noir beckoned with her over-ripe charms, I gave in. Hello Anaïs, Pigalle and the café Wepler where Henry Miller met the beautiful whore Nys – it couldn't be by chance that the name of the sensuous, ravenous, sweet-tempered tart of *Quiet Days in Clichy* was encased in my own … Narcisse Noir's opulent white bouquet harked back to my very first Parisian perfume, Chloé; it prefigured a continuing obsession with narcotic, milky, femme fatale blossoms. After the tenderness of Farnesiana, the toughness of Poivre and the sultriness of Narcisse Noir felt like declarations of independence.

I'd opted for sexual monogamy but, clearly, I was no longer capable of being as faithful to a fragrance as I had been to

Habanita in my more adventurous days. My senses craved the variety once afforded by the skins and smells and fantasies of the men I'd known, the places I used to zip off to on a whim because I didn't have to hold down a job, the personas I used to try on as I moved through different cities and social circles. Perfumes were a low-risk substitute for those adventures, genies quietly waiting for my summons in their crystal bottles … Fidelity had never been my forte.

16

'No matter what the weather, rain or shine, it's my habit every evening at about five o'clock to take a walk around the Palais Royal,' writes Denis Diderot in *Rameau's Nephew*. 'I let my spirit roam at will, allowing it to follow the first idea, wise or foolish, which presents itself, just as we see our dissolute young men on Foy's Walk following in the footsteps of a prostitute with a smiling face, an inviting air, and a turned-up nose, then leaving her for another, going after all of them and sticking to none. For me, my thoughts are my prostitutes.'

It was in the very spot where Diderot had indulged in his intellectual *libertinage* that I found the scented seraglio that best suited my fancy. The Palais-Royal, a garden in the heart of Paris so secluded it is missed by most tourists, was where I went to speak to the witty, amiable ghosts of the 18th century. It was teeming with them, the rakes and the courtesans, the philosophers and the coquettes, the aristocrats and agitators, floating in the honeyed scent of linden tree blossoms, magnolias and hyacinths.

In 1992, after a long spell away from my favourite haunt, I spotted a new boutique with a purple façade and black windows in the Galerie de Valois that looked as though it had popped out of some wormhole connected to the 1780s. This was the kind of shop Casanova might have opened to trick elderly aristocrats into thinking he was a powerful cabbalist. 'Shiseido Salons du Palais-Royal' said the storefront. I pushed the heavy glass door and wandered into the dark, cool shop. Stylized lavender astronomical motifs adorned the deep purple walls; a delicate spiral staircase ascended to the first floor. A sphinx-like sales attendant clad in a purple smock like those worn by the staff of great jewellers stood behind a marble counter, where four bell-shaped glass bottles were set next to a miniature 19th-century lion-footed marble bathtub in the Pompeian style.

If the place hadn't been so dauntingly quiet, I would have squealed when I learned who was behind this new perfume house: the man whose mysterious, violently geometric pictures of Kabuki-faced women with smoky cat eyes and thin, cruel scarlet lips against half-Japanese, half-Russian Constructivist backgrounds had fascinated me in my French neighbour's *Vogues*. His ads for Christian Dior cosmetics had driven me to buy my first red lipstick on my first trip in Paris and when he'd moved to Shiseido, I'd shifted allegiance as well. But somehow, the fact that he made perfumes had passed me by. It was better this way; to discover them as though guided there by my ghosts …

I had just entered the world of Serge Lutens.

Those bell-shaped bottles held liquid emotion. Letting Bois de Violette's amethyst and umber tones unfurl from my wrist as I wandered under the arcades of the Galerie de Valois, I couldn't quite decide whether I liked it or not, so unusual was its blend of oily, resinous, leathery woods inlaid with sweet shards of violets and bits of golden dried fruit. But I knew I loved it as you would a stranger who seemed to carry with him the mystery of his own world. And there *was* a world behind it; there was a story. Just as I, a woman from the snow country, had come alive to scents in Seville, Serge Lutens, born in the northern French city of Lille, had discovered the olfactory realm in Marrakech, where he'd settled in the late 60s.

In an industry choking with too many launches and where fragrance had become a consumer product, the house of Serge Lutens rang out as a protest, a sovereign gesture of defiance: *Qui m'aime me suive*, whoever loves me shall follow me. It was raised on the foundation of his aesthetics and his persona – his 'personal legend', as Paulo Coelho would say, though I suspect he is no more Lutens' favourite writer than he is mine. The fragrances were only available in a single shop, their discovery a ritual experience. The sophisticated stage Lutens set down revived the mystical couture atmosphere he had discovered when he came to Paris to work for *Vogue* in 1962.

At the core of Serge Lutens' stance was a violent rejection of the mainstream in general but more specifically of the streamlined, limpid style established by Edmond Roudnitska, the influential composer of the best-selling Diorissimo, Eau Sauvage and Diorella. Serge Lutens knew these well, since he'd been the artistic director of Christian Dior's makeup line from 1967 to 1980. 'With him, it is the start of a cleaned-up perfumery, with neither body nor memory, prim and proper and which awakens nothing in me. Perfumes have to belong to our roots, our sweat, our past and our very decadence,' he explained to Annick Le Guérer in *Le Parfum*.

To me, Lutens' reintroduction of an archaic dimension in perfumery echoed the gesture of the Italian director Pier Paolo Pasolini, unsurprisingly one of his favourite directors. When Pasolini adapted Greek myths and tragedies (*Oedipus, Medea*) and pre-modern narrative cycles (*The Thousand and One Nights, The Decameron, The Canterbury Tales*), he sought to restore the 'primitive' gaze; to show the world as it was seen *before* Christianity or the Enlightenment. Lutens' perfumes, with their vibrant palette of oriental aromas, were a similar move towards pre-modern conceptions of the art, albeit with the financial support of the Japanese cosmetics giant Shiseido.

In keeping with Lutens' quest for the ancient roots of perfumery, the house quickly established, though he denies it – 'they just *came*' – its Moroccan-inspired olfactory codes: spices, predominantly cumin, dried fruit and the rich, leathery Atlas cedarwood. But Lutens' style went beyond an exotic palette. He often asked his perfumers to take on difficult notes that had seldom or never had the starring role in mainstream perfumery, and to exaggerate their characteristics to the point of distortion. Iris Silver Mist, for instance, was the first iris soliflore in decades, namely because orris butter, which takes years to produce, is extravagantly expensive. It is mainly used for its powdery effect, but it also smells of wood, roots, carrots and earth, with cold, metallic effects and fatty whiffs of human flesh. The perfumer Maurice Roucel, egged on by Lutens, boosted those facets and went on to produce one of the most austere perfumes on the market.

Muscs Koublaï Khan lurches in such a different direction you'd think it had been thought up by an entirely different author (and it was, indeed, composed by another perfumer, Christopher Sheldrake). But, in fact, it has the Lutens signature stamped all over it. Just as Iris Silver Mist over-saturates all the facets of iris including the less flattering ones, Muscs Koublai Khan piles layer upon layer of animal notes to achieve a rendition of the legendary Tonkin musk encountered by Marco Polo in the Mongol emperor's Chinese realm, hence its name. To some, it is one of the fiercest stenches ever to waft from a perfume bottle, and it does feature a cornucopia of feral smells: faecal civet, leather and fur-smelling castoreum, costus with its whiff of dirty hair, armpit-reeking cumin, ambergris with its saline, female notes, patchouli and its dank earth facets, as well as a wide range of synthetic musks. But despite this reverse-laundry list of pungent materials, Muscs Koublaï Khan doesn't add up to a devil's brew. In fact, it may be the fragrant equivalent of Ingres' *Turkish Bath* as described by Kenneth Clark in *The Nude: A Study in Ideal Form*: 'In the middle of this whirlpool of carnality is [Ingres'] old symbol of peaceful fulfilment, the back of the Baigneuse de Valpinçon. Without her tranquil form, the whole composition might have made us feel slightly seasick. [But] after a minute we become aware of a design so densely organized that we derive from it the same intellectual satisfaction as is provided by Poussin and Picasso.' In MKK, the figure of the 'baigneuse' is the crystalline rose and ambrette accord: a note of tranquil harmony rising through the animalic arabesques.

From the time I discovered the Salons du Palais-Royal, the bell-shaped bottles lined up on my bedroom mantelpiece, each new scent a yet-undiscovered room in Lutens' olfactory palace. It would be nearly ten years before I met the man himself – a slender, sloe-eyed sylph with the light step of a dancer and a deep, velvety voice. That day he said:

'We've met before, haven't we?'

We hadn't, I assured him.

'Somehow, I'm sure I know you.'

'Well … in another life, then?'

17

As soon as I am ushered up to the private room above the Salons du Palais-Royal, done in saffron and cumin in contrast to the mauve and black of the ground floor, Serge Lutens insists: 'It's funny, when I first met you, I thought I knew you. And today, when I saw you, I told myself again I know this person, but I don't know why.'

This uncanny sense of familiarity was no shortcut to obtaining an audience. When at long last I had been invited to one of his launches, I had introduced myself, gushing in a most un-Parisian fashion that I'd admired him since the age of twelve. He'd looked at me as though he had known me for that long too, and after taking leave of me had doubled back on his steps to tell me that when he said *au revoir*, it wasn't a mere manner of speaking: we *would* see each other again. But when I took him at his word and put in a formal request for an interview, I was asked to forward my questions in writing along with a covering letter. I described myself as he'd seen me: 'Silver hair, scarlet lips, black trench coat.' He answered that I defined my silhouette as a battledress meant, perhaps, to protect myself, and that this may have been what had made him come to me intuitively, without a word.

I'd been led to understand that we would elaborate on his written answers over the course of our face-to-face interview but, as it turns out, Serge Lutens isn't much interested in discussing them. Nor am I, for that matter. I haven't come for answers. I've come to see the Wizard.

A mutual acquaintance had told me that he might enquire about my astrological sign – he himself is Pisces with Leo rising – and he does. If he asks, he explains, it is out of shyness and curiosity, to get to know people a little more quickly. I am Aries with Cancer rising, I tell him, which sets us off on a discussion of the hot Arian temper, which he advises me not to rein in. But I'm not one to display my anger:

'Anger is very intimate, and sometimes you don't want that intimacy.'

'We don't want to squander our anger on just anyone, do we?' he quips.

'So what about you?'

'I can be very choleric, but ridiculously! I'm completely infantile … I lose it in front of everyone. I'm a disaster. My angers are frankly … irrational. After that, I'm sorry.'

This isn't an interview, the ritual in which one person extorts as much as she can from another without disclosing anything. Lutens is asking *me* questions, a sly way of wiggling out of mine about his perfumes, I suspect. He says, once he's done with them, he's through. So we speak in circles around them as his assistant pours him tea and he pulls out his round glasses from his pocket without ever putting them on.

From astrology we slide into tarot. His photographs have always seemed to me to be somewhat tarot-like figures, their meaning endlessly reversible. Lutens doesn't read the tarot. He doesn't even play cards: 'I'd be sure to lose.' But he does know something about the reversibility of signs and sentiments. For instance, he seems to have taken a perverse pleasure in launching a 'clean laundry' scent that ran counter to every expectation, L'Eau Serge Lutens. His fandom resonated with cries of treason, I tell him. The word amuses him.

'Would you enjoy being seen in the same way after twenty years? You tell yourself that you'll be better understood as time goes by, but also that you'll disappoint more and more. This can be called treason, but it's not yourself you're betraying.'

It's no wonder Jean Genet, the homoerotic thief who turned treason into the driving principle of his life, is Lutens' favourite author: 'Anyone who hasn't experienced the ecstasy of betrayal knows nothing about ecstasy at all,' Genet writes in *The Prisoner of Love*.

It is then that, in a way, I commit my own act of treason by telling him about my project with Bertrand, which I've never spoken about to anyone in the industry. My disclosure arouses no more than the slight interest one bestows on a mildly interesting piece of gossip, but it's given me a jolt. I've just realized that what I want to learn from Lutens is the secret of the peculiar mind-meld between the one who brings the story and the one who translates it into a perfume formula … Lutens doesn't compose his scents: 'I am not a perfumer,' he says. 'I make perfumes, or rather, I make them talk, or even confess.' For the past twenty years, with one exception, the actual composition has been carried out by Christopher Sheldrake.

Is 'Serge Lutens' only Lutens, or him and Sheldrake? The two men couldn't be more different: Sheldrake is a sweet, straightforward family man who, says Lutens, likes food and wine, lives in a light-filled house and dresses in neutral colours. By contrast, 'I love dark houses, strong perfumes – the ones that leave traces. Black is their colour. Eating and drinking are pointless to me and my solitude is rich.' Their aesthetics don't complete or influence each other, he tells me: they 'listen to each other'.

'During its gestation, perfume is organic; it moves by itself and within itself, and sometimes leads us to unpredictable and unsuspected paths. If I were to say it leads us by the nose, it wouldn't be false. Day after day, it explains to me who it is and what it is. Once it's done, I give it a name and follow it until it has become what it needs to become. Then we part ways, never to see each other again, and I move on to the next one. It's like a passing fling ... almost nymphomania!'

Mr Lutens may come off as an otherworldly sprite, but he's actually quite a funny man. As we burst out laughing, I find myself nodding vigorously. He hasn't disclosed a thing about the way he works with his partner-in-perfume, but he's told me what I need to know: to follow the perfume 'until it becomes what it needs to become'. Duende must dictate its own terms – the last thing I want is a bespoke perfume. I've never seen the point of them. As for Lutens, he thinks they're a joke: 'Perfume is made-to-measure *by definition* if you recognize yourself in it. Perfume is the subject, just as the tiger is the tamer's subject. Making a perfume *for* someone? The perfume is forgotten! It's not itself any more!'

He's been asked a hundred times to do it, he adds, 'But I'm not a psychoanalyst! And besides, who'd be able to tell it was a bespoke perfume? You might just as well wear the bottle on a chain around your neck, with a label stating the cost. I am accountable to the perfume, and nothing else.'

Our conversation meanders from his memories of coming to Paris as a young man and discovering the rites and mystique of the great couture houses to the books he is reading – he says his scents are now more inspired by literature than by the smells of Morocco. I pull out a quote by Genet I'd copied down in my notebook: 'Solitude is not given to me, I earned it. I am led towards it by a concern for beauty. I want to define myself, delineate my contours, escape from confusion, order myself.' Of course, he knows it. It could be the motto of a man who spends most of his days reading, and who presided for three decades over the unending embellishment of a palace in Marrakech: 'It was an image I pursued, but I can't live in it, I need disorder,' he explains, which conjures visions of the impeccably turned-out Serge in a djellabah, sprawled on piles of cushions, surrounded by open books, perhaps unshaven and tousled-headed ...

The palace was shrouded in utmost secrecy as long as it wasn't completed: since then, Lutens has allowed pictures of some of the rooms to be published by *W* magazine. I was interested to find out he'd decorated an entire wall with Berber fibulas, etched silver clasps used to fasten traditional garments. It just so happens that I own such a fibula: it is my favourite piece of jewellery and, more than that, my talisman. I'm wearing it today, hung from a slim platinum torque. Its peculiar shape is what attracted me to it: a lozenge ending with a sharp spike, which makes it both a dagger and a shield, a feminine and masculine symbol. It was the first thing Lutens noticed when I came in. Now I tell him how I came to buy it. It was, of course, in Marrakech.

I made it to Lutensian Ground Zero in 1999. I'd gone with the Tomcat to do a story on the last survivors of the Yves Saint Laurent posse, ensconced in the riads they had transformed into *One Thousand and One Nights* palaces with the help of peerless Moroccan craftsmen. I knew Lutens had been working on his own house and sent out feelers to secure an introduction. But he didn't participate in the endless round of dinners and fêtes of the expat community, at least not the ones I knew. When I spoke to an architect who'd been consulted on it, he said he was sworn to secrecy. Fair enough. Besides, I had other things on my mind.

From the instant I'd set foot within the pink walls of the old city, its winding alleys that seemed as though they had been secreted by the earth itself giving way to cool inner gardens choked with gaudy-coloured bougainvilleas around cooing fountains, I'd felt at home, my senses in overdrive. It was Seville all over again, as it must've been under the reign of the Caliphs. The Koutoubia minaret loomed, identical to that of the cathedral of Seville, over the central plaza of Djema-el-Fnah with its storytellers, charms vendors, teeth-pullers crouching behind piles of yellowed molars, carts of mint pulled by donkeys and pyramids of oranges to be squeezed for fresh juice ...

And it was spring. And I caught it with a pang, wafting from a garden, and I tumbled back in time to the woman I'd once been, to Carmen. It had been building up, this need to throw off my shackles. My impatience with the Tomcat, with his expectation that I would set aside my literary ambitions while he wrote 'the Novel' or 'the Script' ... The Neverland of Marrakech, the suspended reality its inhabitants had built for themselves away from the West, had revived an ache to live in beauty all the perfume collections in the world couldn't have soothed, not even Serge's alchemical potions.

I stood at the threshold of the garden, abstracted. Then I stepped into the garden and saw the orange trees in blossom. And when I came back, as Leonard Cohen sings in 'Famous Blue Raincoat', I was nobody's wife.

Before we part, I ask Serge Lutens one last question. Which of his perfumes he would pick out for me? His verdict is swift:

'The one that condemns you.'

18

To mark the start of our time together, Monsieur had bought a new watch for himself and one for me. Then he'd asked me to find us each a new fragrance. He'd only specified he wanted something with lavender. It didn't take me long to unearth the little-known Mouchoir de Monsieur by Guerlain, launched in 1904 and kept in production for Jean-Claude Brialy, a French actor who'd gone from Nouvelle Vague icon to society darling. The epicene notes of the 'gentleman's handkerchief' – flowers and vanilla as a yin counterpoint to the citrus, lavender and civet yang – suited Brialy's witty, endearing persona, but they also seemed like a good fit for Monsieur, a man whose virile appearance belied the almost feminine delicacy he seldom displayed.

It was this contrast between Monsieur's hearty self-assurance and his utter sensitivity to the finest nuances of my moods that had drawn me to him; that, and the fact that he was more fun than anyone I'd ever slept with: fun to laugh with, fun to play with. It also helped that he could afford to whisk me off to Rome or Avignon on the trail of Caron's Farnesiana and the eponymous Comme des Garçons, treat me to the finest restaurants and bed me in four-star hotels: money made everything lighter. It allowed us to improvise our stolen moments. Monsieur was married. I hadn't divorced yet. We were making the story up as we went along.

Finding my own perfume to herald our clandestine love affair wasn't much of a quandary either. As soon as he smelled Serge Lutens' Tubéreuse Criminelle between my breasts, Monsieur whispered 'Criminelle' in gloating tones and proceeded to bare them.

Tuberose became the olfactory thread of our six-year romance. Flamboyant, assertive, provocatively feminine, it burst with the self-confidence of a woman wholly desired. Its scent was one I had to live up to, in the same way that, amidst the complications and dissembling of an adulterous affair, I felt an almost ethical – or was that aesthetic? – urge to keep things light and playful. Wasn't that the whole point of taking a lover? Somehow tuberose embodied this love-as-aesthetic-performance stance: a cold, venomous green bitch-slap subsumed in narcotic, creamy, coconut-white flesh betraying hints of a rubber soul.

The couturier Robert Piguet sussed the diva out perfectly when he dressed up the first best-selling tuberose perfume in black and hot pink. The aptly named Fracas ups the exuberance of the flower until it reaches the shrill peaks of a soprano coloratura: *The Magic Flute*'s Queen of the Night, with her dramatic entrance and near-hysteric trills. Shades of Germaine Greer's *Female Eunuch*, Fracas could very well be the perfume equivalent of a Hollywood persona accumulating the signs of femininity on a body that is already feminine (Marilyn Monroe) or not (Ru Paul). Fittingly, Fracas is said to have been the signature fragrance of Madonna (the ultimate manipulator of feminine semiotics), the late fashion editor and muse Isabella Blow and the resplend-ent beauty whose modelling career she launched, Sophie Dahl: this is nothing if not the scent of divas.

But much as I admire Fracas, it resists me. In fact, it lived up to its name by shattering on my floor as soon as I took my vintage bottle out of its box. It was the weirder Tubéreuse Criminelle that introduced me to the dark side of the flower and set off my addiction.

When Serge Lutens and Christopher Sheldrake decided to tackle tuberose, rather than attempting to tame the jarring medicinal notes of tuberose absolute, they decided to amp them up. The result is the olfactory equivalent of an extreme close-up: the features of the tuberose are so distorted they are barely recognizable, as shocking as the face of a woman in the throes of pleasure seen at kissing distance. But this floral monstrosity fleshes out into an intoxicating potion: the *belle-laide* awakens sensations that a merely pretty girl could never hope to achieve …

In the language of flowers, the tuberose speaks of dangerous, forbidden pleasures: by taking up the Criminelle as the emblem of my passion, I'd guessed that message in a bottle. Perfumes are our subconscious. They *read* us more revealingly than any other choice of adornment, perhaps because their very invisibility deludes us into thinking we can get away with the message they carry, olfactory purloined letters lying right under everyone's nose … And perfumes *themselves* have subconscious, twisted molecular secrets sending out subliminal calls of the wild. Don't accuse perfumers of having a dirty mind, though their peculiar variety of sensual gluttony often *does* draw them to louche aromas.

Put the blame on Mame, boys.

Tuberose and her sisters jasmine, orange blossom, gardenia, honeysuckle are the vamps of the floral realm, pallid creatures whose hypnotic, diffusive scents are potions for attracting nocturnal pollinating insects – vividly coloured flowers need none to draw bees in the daytime. Their velvety flesh feels like a woman's skin and, even at their freshest, a hint of corruption wafts up from their sweet fragrances. Stick your nose in them. Go past the pretty. Zero in on the weird. Butter, Camembert, mushrooms, horse manure, bad breath, dirty feet, blood, meat, shit … Despite their tiny size and pristine petals, white flowers bellow Nature's obscene secret through their outsized fragrance: flowers are sexual organs. And if those sexual organs have ended up grafted David Cronenberg-style onto our skin, it is precisely *because* they also smell like the human body in all its extreme states, whether pleasure or death.

And therein lies the secret of their sexiness, if we are to believe Dr Paul Jellinek, who wrote *The Psychological Basis of Perfumery* in 1951. The German chemist and perfumer was convinced that the purpose of perfumery was 'to create and enhance sexual attraction', and classified raw materials according to their erotic effect: anti-erogenous (or refreshing), stimulating, narcotic or erogenous. The latter category encompassed 'perfume materials reminiscent of human body odour' emanating from the scalp, underarms, pubes, urogenital and anal regions. Though repulsive when smelled out of context, these subtle hints of the naked body were most frequently encountered in the act of love and therefore carried positive connotations, claimed Dr Jellinek. He further classified his erogenous materials based on the smells of blondes (sour-cheesy), redheads (pungent-burnt) and brunettes (sweetish-rancid). One can only wonder at the method he used to collect these observations. I envision rows of naked women sniffed by a formal gentleman in a white lab-coat dictating to an assistant in cat's-eye glasses, a scene Stanley Kubrick could've filmed circa *Dr Strangelove* … But though the progress of organic chemistry has rendered many of Dr Jellinek's analyses obsolete, his take on the human facets of raw materials makes for a fascinating read. For instance, costus, the stuff Bertrand thought might be my 'referent', 'is strongly reminiscent of the odour of the scalp region and … of the axillar odour of brunettes.' Score one for Bertrand, then: though my hair has gone prematurely white, I *am* a brunette. Incense also has 'something of the sweet-acrid effect' of brunette sweat. It seems like we're really on to something with Duende … As for orange blossom, what makes it so enticing is the indole it shares with other white flowers such as jasmine and tuberose: 'It is precisely the odour of indole, reminiscent of decay and faeces, that lends [them] that putrid-sweet, sultry-intoxicating nuance which has led to the use of … their extracts as delicate aphrodisiacs.'

I'm quite happy to follow the good doctor's ghost down that path: I *do* believe that reminiscences of a not-quite-surgically-scrubbed body can act as subconsciously enticing reminders of our animal nature. But I'm not quite sure the fatal attraction of white flowers boils down to indoles. There must be more to the story, and to find out, Octavian Coifan is the man I turn to.

Fortunately for me, the erudite author of *1000 Fragrances* is not only a friend but practically a neighbour. I've spent many a Sunday afternoon with Octavian, rows of phials and blotters lined up between us on my dining-room table, parsing raw materials, vintage finds and new products while batting away the cat, who likes to snatch away our blotters. Octavian was the first to walk me through the maze of perfume composition and initiate me into raw materials. When I blurt out 'Am I crazy, or is there such or such a note in this?' he's fond of answering, with a glint in his big green Byzantine eyes, 'The nose is *never* crazy,' which leaves the question of my own sanity open to conjecture …

To find why our skin loves white flowers so much, Octavian takes me on a tour of their chemical plant. Their specifically floral character comes from methyl anthranilate, which basically smells like orange blossom, often associated with benzyl acetate, which bridges the gap between nail polish, banana and jasmine. The spicy notes are eugenols: white flowers share them with cloves, carnations, ylang-ylang and lilies. And then there are the infamous animal notes. Indole, a consequence of the degradation of protein, hence its presence in corpses and faeces, but also paracresol, which is reminiscent of horse manure – unsurprisingly, since horse manure and horse sweat *do* contain the molecule. This is why narcissus absolute gives off a horsy whiff and why a jasmine bush at a certain distance will make you wonder where the stables are. Tuberose also contains a touch of skatol (as in 'scatological') and butyric acid (from the Greek for 'butter', but the effect is cheese and feet). Jasmine even contains compounds similar to those found in tobacco, which goes quite a way to explaining why the heady floral perfumes worn by the femme fatales of yore blended so divinely with the blue wisps of their cigarette smoke – picture Bacall growling 'Anybody got a match?' to Bogart in *To Have and Have Not*. Her nickname may have been Slim, but when Bogart had a surreptitious sniff of her perfume bottle, it wasn't wafting some shower-clean anorexic juice.

Smell jasmine, tuberose or gardenia and you'll also pick up a fatty/buttery smell with coconut, hay and peach facets. These are produced by molecules called lactones, from the Latin for 'milk'. Gamma-nonalactone, commonly known as aldehyde C18, is what gives Hawaiian Tropic-type tanning products their characteristic coconut fragrance. Another, C14 (gamma-undecalactone) smells of peach. Certain types of musks, such as the ones naturally found in angelica and ambrette (hibiscus seeds), are lactones too. It just so happens that, like flowers, we are lactone factories: they are produced on our scalps when a certain type of bacteria feeds on sebum. Or rather, they would be if we didn't lather daily. When Charles Baudelaire wrote, 'In the downy edges of your curling tresses/I ardently get drunk with the mingled odours/Of oil of coconut, of musk and tar' in a poem aptly entitled 'Head of Hair', he was in fact exercising a keenly analytical nose in the midst of erotic rapture.

My particular favourite, tuberose, is the white flower that contains the greatest quantity and variety of lactones, unlike Duende's orange blossom, which has none: this is one of the reasons why Bertrand told me that it was a 'hard' smell. But if natural jasmine oil is most reminiscent of skin, Octavian explains, it is also because of the several dozen molecules that don't play a prominent role in its smell and therefore aren't used in synthetic reproductions of the notes. Esters of fatty acids, for instance: jasmine is a member of the *Oleaceae* family, just like the olive tree, and contains them more abundantly than any other flowers … just like our skin, where those substances are also present.

It is the combination of the decaying smell of indoles with the skin-and-scalp fattiness of lactones and esters of fatty acids that weds the scent of white flowers to flesh and pulls them halfway into the animal kingdom, which Octavian confirms by two counter-examples. Lily-of-the-valley is one of the most indolic flowers but it isn't perceived as sensual because it has neither lactones nor spicy molecules; the scent of chestnut blossoms is mostly made up of indoles and molecules containing nitrogen, but similarly devoid of lactones.

'And they smell like sperm!' I blurt out, recalling the embarrassingly vivid scent that spreads over the neighbouring avenues in spring. 'Not quite what you'd want to squirt on skin, at least not from a perfume bottle.'

Octavian tells me that in his native country, Romania, this is also what the peasants say. I pull out a short story by the Marquis de Sade bearing on the very same subject: a young girl innocently exclaims she recognizes the smell of chestnut blossoms, much to the embarrassment of her mother and her confessor … Nature definitely has a one-track mind.

When you smell white flowers, 'you smell notes that are part of human life from birth to death and on to decomposition,' enthuses Octavian. Even when they start out with a cut-grass smell as cool as a rain-drenched English garden, they're also perversely giving off whiffs of decay, and it is this contrast that gives them such depth. Smell a honeysuckle hedge at night, when the flowers are not so fresh, and you'll get the smell of a human presence, of human breath. Even after they are picked, they go on producing their scent. They are *dying*, not fading: 'The white flower is a flower that decomposes in sheer beauty!' my friend concludes.

We stop to ponder the fascinating continuum of Nature, which bathes vegetal and animal flesh in the same odours:

perhaps wearing perfume is our way of reaffirming this atavistic bond? Their oddly compelling blend of erotic attraction and death truly does make white flowers the femme fatales of perfumery. It might also explain why they elicit mixed feelings in certain cultures. The Aztecs used to pile them up on their pyramids, where their smell blended with the stench of their sacrificial victim's blood – there is a bit of a blood note in the tuberose, and that hint of blood is, again, a bond between death and life, human sacrifice and female fertility, Octavian tells me. In Malaysia gardenias are used exclusively for funerals, I recall reading; lilies are not uncommon ornaments in funeral homes in the US, which is why some Americans have an aversion to lily-centred scents. Octavian read in a trade magazine from the late 30s that tuberose was rejected in the US for that very same reason. At a time when embalming techniques weren't perfect, the flowers' indoles mixed with the corpses' and this boosted the scent of the flowers while masking the stench of death. Win-win.

Oddly enough, tuberose fragrances went on to become highly successful in America: you can lay even money that when a brand comes out with one, it is to court the dollars. This brings us to what I'd call the 'white flower paradox'.

Why were there so few white floral compositions, we wonder, during the intensely creative era spanning from the 1910s to the 1920s, when most of the great templates of perfumery were invented? After all, jasmine, tuberose and orange blossom had been mainstays of the perfumers' palette for centuries and their smell was just as sexy as those of the leather, vanilla and resins that were so fashionable at the time. It's not that there weren't any. Though tuberose perfumes were sparse after the late 19th century, there wasn't a perfume house that didn't have a jasmine or a gardenia (the fragrance of gardenia can't be extracted but that didn't keep perfumers from approximating it), including the modernist Chanel, though she'd claimed a woman shouldn't smell like a flower. But none is known as a milestone of perfumery. Was it because, in a period marked by the invention of great abstract fragrances, a plain old floral seemed distinctly unexciting to perfumers and their clientele? Too reminiscent of the naturalistic blends of the 19th century? The reason, Octavian ventures, may have simply been cost: from World War I onwards, French-grown jasmine and tuberose became so expensive that they couldn't be used in large quantities. But if perfumers had truly wanted to reconstitute the scent of tuberose by using a blend of less costly natural and synthetic materials, they could have. Everything was available in the 20s, and the major white floral bases that were used up to the 80s were indeed invented during that decade. Jean Patou's Joy, launched in 1929 as a gesture to his American clients when the Wall Street Crash prevented them from crossing the Atlantic to visit his salon, was overdosed with jasmine, but also rose, and therefore not strictly a white floral. But there *was* a white floral scent that became a best-seller as far back as 1911 I remind Octavian: my old Narcisse Noir.

'Yes, but on the American market!' Octavian shoots back.

'And what about Fracas? That came out in the 40s and it must've been a blockbuster since it managed to survive the demise of the house of Piguet and several changes of hand over the decades,' I insist.

'On the *American* market!'

True. Between the both of us, we can't name another best-selling white floral before the late 70s. Karl Lagerfeld's Chloé and Cacharel's Anaïs Anaïs were the first waves of what would become a bona fide tsunami in the 80s, starting with the gaudy – and American – Giorgio Beverly Hills in 1981, followed by Givenchy's Ysatis, Guerlain's Jardins de Bagatelle and Dior's Poison. None was purely a white floral, but all boosted their tuberose, orange blossom and jasmine notes to unprecedented intensities.

So, what happened between the late 70s and the mid-80s, we muse …

'The American market?' I venture.

It makes sense. After Opium, which had raised the stakes both in terms of intensity and planetary success, the perfume industry needed products that felt modern – modern in the 80s being vibrant colours, sequins, quarterback shoulders and helmet hair. Products that smelled as brash as Alexis Carrington making a play for Gordon Gekko. Products that gave out a message loud enough to be heard around the world. The intellectual chypre family epitomized by Diorella or cashmere-and-pearls floral aldehydics like First, both expressions of genres invented half a century earlier, were too complex, too classy, too *French* to make the cut. Even the perennially best-selling N°5 came off like Grandma's perfume all of a sudden. Whereas tuberose and her sisters were *out there* – however I love them, one thing they are *not* is subtle. They are entrance-making, resistance-is-futile perfumes that'll tattoo your presence all over a room, not to mention a gentleman's body. Throw me on that couch and ravage me, or else you'll be sorry. At least that's the implication, even though any attempt in that direction would probably trigger a swift, spike-heeled kick in the baubles.

'So … do you think it's my North American side that makes me love tuberose so much?' I ask Octavian a bit anxiously – I pride myself on my sophisticated Parisian tastes.

He gives me one of his draw-your-own-conclusions smiles.

'Mind you, I never wore any of those perfumes in the 80s.'

'You love Poison, don't you?'

'I do *now*. Back then, I was too much of a snob to wear something that popular.'

'Women are so complicated,' he sighs. 'See what you missed out on all those years?'

Since then, I've caught up with a vengeance. I'm not much of a one for jasmine if it's not on the bush; orange blossom starts interesting me when it's a little messed up; gardenia only suits me when it's as decadent as Tom Ford's now-discontinued Velvet Gardenia, which exaggerated the smelliest traits of the flower.

But I am the Empress of Tuberose, and the perfume that confirmed the addiction sparked off by Tubéreuse Criminelle was the equally aptly named Carnal Flower.

Ironically, just as the Lutens had sounded the opening notes of my love affair with Monsieur, Carnal Flower, unbeknownst to either of us, was its coda. I'd come to one of our lunches in Saint-Germain-des-Prés after dousing myself from the tester. With its eucalyptus and ozonic top notes boosting the tuberose's minty-camphoraceous bite, Carnal Flower seemed to carry the coolness of the florist's storage room, just as the blistering air of January was caught in the hairs of my black fur wrap. Its voluptuous creaminess and musk expressed the warmth of the flesh under the fur. Smelling it was like diving in slow motion into the flower, every nuance magnified as though it had been dissected, analysed and reassembled in different proportions by an artist-engineer who'd second-guessed Nature … Monsieur adored both the scent and the name. After lunch, he dropped me off in front of the shop on the rue de Grenelle, practically opposite the Christian Louboutin boutique I'd so often pillaged in his company, clutching a two-hundred-euro banknote – he couldn't find a parking space and he was late for his next appointment so, if I didn't mind, I'd have to make the purchase myself.

I'd told Monsieur from the start I wouldn't be his mistress for more than five or six years. In the meantime I'd divorced, though not for Monsieur: I'd already decided to leave the Tomcat when Monsieur and I met. He hadn't. I'd never asked him to. And all of a sudden, it was time to end it. I'd made Monsieur happy and that happiness had made him more successful than ever in his profession; success meant he was no longer free to abscond with me for a few days. There were more lunches and fewer trips. Like a tuberose, our affair was fading beautifully, but I wasn't going to wait for it to turn into compost. So whenever he called I wouldn't be available and sometimes it was true.

But I didn't know that Monsieur would never get to smell Carnal Flower on me again when I pushed the door of the red and black shop, banknote crumpled in an elbow-length black leather glove, giddy with champagne and a little wobbly on my Louboutins. I'd found an alternative to Tubéreuse Criminelle, a new incarnation of the flower, and the perspective of a tryst fired me up.

I'd started cheating on Serge, you see. I had been for quite some time – in fact, just about when I'd started cheating on my husband. His name was Frédéric Malle.

19

Remember the scene towards the end of *Manhattan* when Woody Allen muses on what makes life worth living? He goes on to list Groucho Marx, Louis Armstrong's recording of 'Potato Head Blues', *Sentimental Education* by Flaubert, 'those incredible apples and pears by Cézanne …'

For me? Tintoretto's 'Lady Baring Her Breasts' at the Prado; Marcello Mastroianni's gaze; Serge Gainsbourg's 'Sous le soleil exactement'; the first page of Diderot's *Jacques the Fatalist*; The Ronettes' 'Be my baby' …

Edmond Roudnitska's Le Parfum de Thérèse.

There are some fragrances whose balance is so perfect you feel the merest whisper would upset it. Le Parfum de Thérèse is one such fragrance. I reach for it when I need to remind myself that, despite the crass commercialism of the perfume industry, it *did* produce things as unquestionably beautiful as the pages, paintings or songs that give me such joy. The honeyed melon sprinkled with mandarin, bergamot and clove exhaling a tender jasmine breath; the spiced, rounded plum kissed with green tartness; the radiance that keeps unfurling until the dark moss and leather base, anchored to the skin by a warm, creamy base as the jasmine deepens into over-ripe fruit … Each spray sends a shower of sunshine on my skin.

For nearly four decades, Le Parfum de Thérèse was known as La Prune ('The Plum') and was exclusively worn by Thérèse, the wife of Edmond Roudnitska. It was on the verge of being commercialized several times, including by Guy Laroche as Fidji: it was already in the stores when the couturier pulled it to replace it with another product. Such was its rotten luck that the Roudnitskas jokingly gave it a nickname that is the French for 'lame duck', *le pelé, le galeux, le tondu* ('mangy, scabby, shorn'), their son Michel told me.

As a boy, Frédéric Malle had caught wafts of La Prune when the Roudnitskas visited the Dior offices. His grandfather Serge Heftler Louiche, a childhood friend of Christian Dior's, had founded the couturier's perfume house. His mother, Marie-Christine de Sayn Wittgenstein, who worked there for nearly five decades, was responsible for image, development and packaging. When Malle decided to open his own house, he asked Mrs Roudnitska, a widow by then, and Michel, who'd gone on to become a perfumer, if he could put out La Prune. The perfumer Pierre Bourdon, who had been trained by Edmond Roudnitska and whose father had been the deputy director of Dior Parfums, had advised Malle to do so: he considered La Prune to be one of the best perfumes in history. This symbolic gesture would make Malle the third generation in his family to offer a Roudnitska masterpiece to the world. He couldn't imagine starting without it. Deeply moved, Thérèse Roudnitska looked him straight in the eye and answered: 'I've been waiting for you.'

The company Frédéric Malle set up in 2000 is called 'Éditions de Parfums' and it works just like a publishing house. Perfumers are granted creative freedom, with no deadlines or budget restrictions; as their editor, Malle supports them throughout the development process. Their authorship is not only acknowledged but put forward: their pictures, names and mini-bios are printed on the box sleeves as they'd be on a book.

The idea sprung from Malle's dismay at seeing that no one had thought of inviting Édouard Fléchier, who had composed Poison for Dior, to the spectacular launch party hosted by his mother in 1986. Edmond Roudnistka himself had never been invited to a Dior launch. This was the shocking oversight Malle set out to correct. Treating perfumers as authors seemed like such an obvious move that he wondered why it hadn't been done before; he even worried there might be a catch he was unaware of. There wasn't. The concept of 'author perfumery' was a turning point within the industry as well as for the public, who had been unaware that there were people behind the scenes who created the products sold under designer labels – something brands always kept silent about, preferring customers to think perfumes somehow sprang full-blown from the mind of designers.

The idea immediately made sense to me when I crossed the street from Christian Louboutin's rue de Grenelle boutique on the advice of its suave, Franco-Egyptian manager – a boy who could give you the feeling he was sprinkling rose petals at your feet when he leaned down to slip your foot into a shoe. The perfumes I discovered that day had the character and originality of my beloved classics but they felt modern, and that was something I craved. However timeless the beauty of those classics, sticking to them would have been like dressing exclusively in vintage clothing, and sometimes you need to shed the girdle. The Lutens oeuvre had allowed me to do so, but it was exciting to explore other styles: after all, you don't listen to just one composer or stick to one writer. By the same logic, once you've given up on wearing a single, signature fragrance, why not build up a library of scents? A *library*, not a wardrobe of perfumes to suit various moods, occasions or seasons: the distinction is subtle but far-reaching. Malle's clients were discreetly encouraged, if such was their pleasure, to contemplate perfumes as aesthetic objects. Their first contact with the fragrance was no longer via skin or its unsatisfactory substitute the blotter, but through its disembodied, invisible presence in the air of glass columns that looked like futuristic phone booths, though you didn't actually walk into them: you opened a panel and thrust your head inside. The elaborate installation was not just a marketing ploy or a way of keeping the air in the shops untainted: it was the first proper exhibition space for fragrance.

I walked out that day with Une Fleur de Cassie, yet another call from Carmen. Like my old love Farnesiana, it is built around the *Acacia Farnesiana* and it is this flower that Carmen throws to Don José, the hapless officer whom she bewitches, rather than a rose or a carnation, as is commonly imagined …

When I finally meet Dominique Ropion, a senior perfumer at International Flavors & Fragrances, I am the one to tell him about the Carmen connection, much to his surprise and delight. Ropion is credited with over fifty compositions since the mid-80s, including Amarige and Ysatis by Givenchy, Kenzo's Jungle Elephant and Thierry Mugler's Alien, all best-sellers and impressive achievements. But it is Une Fleur de Cassie and Carnal Flower I've come to discuss: in my opinion, they are two of the best perfumes to come out in the past twenty years, and the proof that their author is one of the most brilliant portraitists of flowers in the industry, with an exquisite sense of balance that owes as much to intelligence as it does to taste.

A quiet, sharp-gazed man not given to poetic flights of fancy, Ropion looks as though he'd be more likely to pore over Schrödinger's equation than perfume formulas, and he did in fact get a degree in physics before training as a perfumer. But as a young man, he says, he was more drawn to the arts, and it is both as an artist and as a scientist that he studies the secret harmonies developed by nature. In an interview with Annick Le Guérer for *Le Parfum*, Ropion compares his approach to that of classic painter: 'To unveil the structure, the anatomy of a flower, a fruit, a bud, they would emphasize one aspect rather than another, isolate it from its context, thicken the line, represent it from different angles, go towards the infinitely small, reveal bits that would have otherwise remained invisible.'

This is how he worked on Carnal Flower, teasing out and underlining certain aspects of the tuberose unearthed by the scientists at IFF (International Flavours and Fragrances) during fine-tuned analyses, in order to achieve a rendition that owed nothing to the iconic Fracas, up to then the template of every tuberose perfume. His first efforts produced a strikingly natural-istic hologram, he explains. But it was an odour, not a perfume, a distinction that often crops up in the industry: it means that the product, however realistic its rendition of its model, doesn't connect with skin. This left Ropion with a problem: how to make his tuberose wearable without resorting to the usual tricks of the trade. The tried-and-true notes that had been used for decades would end up making his tuberose perfume-y and spoil both its originality and its natural feel. Ropion subtly tweaked his composition by working on lactones, present in the tuberose, and musks, which aren't, until he reached a balance that would allow the flower to be perfectly recognizable, yet let it blend with flesh …

He worked in the same way on Une Fleur de Cassie, but the end result feels much more complex and abstract than his tuber-ose because, he explains, cassie is a less familiar smell. In fact, he has found it something of an enigma ever since he was a student perfumer. It reminds him of a sculpture he once saw, he tells me, a large bronze sphere with apertures through which you felt you could enter: 'You want to go in but you don't quite know how, or what you're going to find inside.'

Thus, the tiny cassie flower opens up onto a secret parallel world: an artist's re-framing of a section of the olfactory universe. Puddles and mud glinting in the spring sun between the clouds; not just the flowers, but the soil. Cassie smells of many things: wet cardboard, balms, wood, cinnamon, violets and bitter almonds, and those facets are stretched out until each is a char-acter in the drama. Paired with mimosa and the sweet metallic sheen of violet, cassie betrays its leathery nature; indolic jasmine draws out the animal in it, just like Carmen's wiles made José forsake the army, his mother and his fiancée. A slash of cumin dirties it up even further with its hints of human rankness. Yet throughout it remains fresh as spring and powdery sweet, which makes it all the more deceptive.

But though he may draw comparisons from painting, sculpture or architecture to explain his method, Dominique Ropion is wary of pushing analogies too far. Ultimately, like most perfumers, he thinks in purely olfactory terms, he concludes. Jean-Claude Ellena, who contributed four scents to Frédéric Malle's collection before being hired by Hermès, confirms this. In his *Journal of a Perfumer*, he writes that there comes a point when he must divest himself from words, images or memories: 'When I can no longer describe [a smell], when it has a consistency, a depth, a breadth, a thickness, when it becomes tactile, when the only representation I have of it is physical, I can give it shape and create.'

How you represent reality. How you transform it. How you frame a section of it to bring out unheard-of connections between its elements. How old materials can express new effects. How new materials can shed light on classic forms or create novel ones. How to give shape to an idea in a way that hasn't been done before, that *you* haven't done before ...

These are not the questions asked by an artisan, a technician or an industrial designer. These are the questions of an artist, and initiatives such as Frédéric Malle's have afforded perfumers the opportunity of pursuing the answers. Of approaching the art of perfumery on its own terms rather than as a response to a commercial brief meant to express a client's brand identity. It isn't by chance that, when he opened his house, Malle placed himself under the symbolic patronage of Edmond Roudnitska, a man who fought long and hard to get perfume acknowledged as an art form. Just as Michelangelo said, 'We paint with our brains, not with our hands,' Roudnitska insisted that perfumes were primarily works of the mind, hence his intense annoyance when perfumers were called 'noses'.

'Composing perfumes is a means of expression just like painting and music, depending on what the composer draws from his materials, which are neither inferior nor superior to colours or sounds,' he wrote in 1968. 'I haven't said all perfumes are works of art of the first magnitude,' he added ten years later. 'Besides, all don't claim to be, any more than those who compose them claim to be artists, since it seems a few of them deny they are. Art is not an obligation; it is only a beautiful possibility.'

Is perfumery an art? The debate is still open – and because it is predicated on what one means by 'art', which nobody seems to agree on, it may never be settled. Obviously, the thousands of gallons churned out by a billion-dollar industry aren't all worthy of being exhibited at MoMA. But then, neither are most things people hang on their walls. Put colours and shapes on a canvas and you can get Thomas Kinkade, the 'painter of light', sold in shopping malls all over America, or you can get Jean-Michel Basquiat. For its part, the industry doesn't consider it is making art; it certainly doesn't want *perfumers* to think of themselves as artists – that might get them uppity – though it is quite happy to sell the idea to the public.

But the niche houses that have been springing up since the mid-90s, like Éditions de Parfums, have provided venues for exploring Roudnitska's 'beautiful possibility'. And they have often done so by questioning, implicitly or explicitly, the very *definition* of perfume.

20

When I was looking for a replacement for Habanita, I had dismissed L'Artisan Parfumeur as too hippie-ish. And indeed, when Jean Laporte founded his company in 1976, that was its story: amiable fragrances, often named after a single note, for people who weren't quite ready to switch from head-shop patchouli and Jovan's Musk to Opium or First. The names expressed a quintessentially 70s nostalgia for honest, hands-on, handmade pieces that let the materials express themselves. The scents were fairly sophisticated constructions, but the fact that they put forward recognizable notes as opposed to the abstract products of luxury brands seemed like a throwback to pre-industrial days when perfumers offered all-natural blends.

The fact is that, despite or because of its nostalgic aura, L'Artisan Parfumeur was a trailblazer, and the template of what would later become the thriving sector of niche perfumery. In addition to naming perfumes after single notes, a practice that had been almost abandoned after World War II (with the notable exception of the various vetiver-based fragrances) but was revived by the niche brands that came after it, L'Artisan Parfumeur took a series of game-changing initiatives. For the first time in decades, a new perfume house had appeared that wasn't linked to a fashion label. It was also, along with Diptyque, the only perfume house to offer home fragrances (though Diptyque actually started out with fabrics and objects in 1961 before moving on to candles, then fine fragrance). And like Diptyque, L'Artisan Parfumeur was the first to establish stand-alone boutiques, something only historic houses such as Guerlain and Caron could boast of, so that customers could get the full experience in a controlled environment.

The team who succeeded Jean Laporte (he went on to found Maître Parfumeur et Gantier) took similarly innovative options. Marie Dumont, who headed the company from 1990 to 2004, and Pamela Roberts, the creative director from 1992 to 2008, were both industry outsiders. The slim, sharp, decisive Marie had been a journalist and an advertising executive, while the petite, soft-spoken Pamela had worked in art publishing before taking a course at the Parisian perfumery school Cinquième Sens. 'We were rather naïve and innocent,' Pamela Roberts recalls. 'We'd never done this, so we had no preconceived ideas.'

Both women drew from their backgrounds and from L'Artisan Parfumeur's repertoire of figurative scents to re-explore one of the early paths of modern perfumery, epitomized by Guerlain's 1906 Après l'Ondée, an evocation of an Impressionist garden after a rainfall. The trend was already nascent in the mainstream as well. The year Pamela joined the company three ground-breaking products were launched. Angel was based on Thierry Mugler's memories of fairground treats; Féminité du Bois reflected the essence of Marrakech as envisioned by Serge Lutens; L'Eau Parfumée au Thé Vert by Bulgari was Jean-Claude Ellena's stylized rendition of the smells of a tea shop.

These scents severed fragrance from its function as an extension of a female or male persona – the rugged guy, the innocent waif or the femme fatale – to turn it into a thing that was beautiful, interesting and evocative in and of itself. It was a different way of telling stories, but with smells; of looking at the world, but with your nose. Marie Dumont and Pamela Roberts had sniffed out the zeitgeist. From then on, L'Artisan Parfumeur's fragrances would be as literary as they were olfactory.

'We set off on extremely personal ideas,' Marie explains. 'Our editorial line was: if I love this, others will. In the end, what is most personal will be what others can share in the most. Exactly like the novelist tells his story, and this story touches others. Our stories weren't stuck onto the fragrance by the PR department as an afterthought: they were written even before we started working on the juice.'

L'Artisan Parfumeur's scents conjure a place in time: a garden with a fig tree, a flower shop, a hedge in summer. Childhood memories: the 'Je me souviens' ('I remember') coffret was a series of familiar scents to be dabbed on a handkerchief. A trip: the 'Odours stolen by a travelling perfumer' collection. Dzing! by Olivia Giacobetti gradually unfolds all the smells of a circus, the wood of the ring, the horses, the caramels sold during intermission, the suave odour of the panther ... Her 'Sautes d'humeur' ('Moodswings') coffret offered scents for every mood, including negative ones. To connect moods with smells, Pamela used metaphors:

'Anger lights up like you strike a match. So with Olivia, we put sulphur in the top note to express the explosion of anger. The challenge was to make it wearable at the same time.'

'Otherwise, it remains an experimental odour,' adds Marie.

As I speak with the two women, I can't help but think of my own collaboration with Bertrand Duchaufour, who was part of their dream-team along with Jean-Claude Ellena, Olivia Giacobetti and Anne Flipo, later joined by Céline Ellena. Clearly, the ladies had a gift for casting. And I understand better now why my scented Sevillian story aroused Bertrand's interest: he's obviously got fond memories of a period that yielded some of his best-loved earlier creations, from Méchant Loup's hazelnut-scented romp in the woods to the spice-laden Timbuktu ... I prick up my ears when Marie explains how it was her role to make sure the perfumer stuck to the original story, and to determine when to stop – shades of Bertrand's vexing question, 'Do you know what you want?' Like Marie and Pamela when they started out, I'm an outsider with a literary background. And however scant my experience, one thing I have is words …

But it is Christian Astuguevieille who Bertrand credits with giving him the opportunity of creating what he considers as his first truly personal work: Calamus, based on the plant of the same name, which he composed in 2000 for Comme des Garçons.

'No one uses calamus because it's really unbearable,' Bertrand told me. 'It has fantastic positive sides, it smells like a cake right out of the oven and, at the same time, it's sludgy and it stinks like tanned hide you've just pulled out of the water; it almost smells like fish skin!' When presented with the idea, Astuguevieille gave him the green light. 'If I started taking on that kind of challenge, it's thanks to him! He was the only one who dared at the time. He's a pioneer because he's an artist himself. He's got stature.'

Though he is an artist and a designer, words are what drive Christian Astuguevieille in his capacity as the creative director of Comme des Garçons Parfums. Just as the Japanese brand's founder and genius designer Rei Kawakubo displaced every cursor of feminine beauty in the 80s by burying breasts, waists and hips under asymmetrical garments, Astuguevieille has pushed back the limits of what could be called beautiful in a perfume and even what could be considered the *subject* of perfume. In this subversive enterprise, language – the story supplied by the name and/or list of notes – plays as important a part as it does in any gallery-exhibited artwork.

Consider Odeur 71, whose notes list reads like a Surrealist collage or an artist's installation: 'electricity, metal, office, mineral, dust on a hot light bulb, photocopier toner, hot metal, toaster, fountain-pen ink, pencil shavings, the salty taste of a battery, incense, wood, moss, willow, elm, birch, bamboo, hyacinth and lettuce juice'.

Sitting in the blindingly white Parisian showroom of Comme des Garçons, Astuguevieille, a dapper, white-bearded gentleman, tells me how the iconoclastic Odeur 71 kicked off. He'd called IFF to say he'd be coming in to give his brief late in the afternoon. When he got in around 6 p.m., he asked where the photocopier was, dragged the five perfumers who were interested by the brief to it and told them: 'This photocopier has been on since this morning, it's overheated, it smells, and that smell is your starting point. That's my brief. We're going to work on overheated objects from our everyday life.'

Perfumes such as Odeur 71 focus on 'found smells' much as artists produce ready-mades by pulling mundane objects out of context to exhibit them. L'Artisan Parfumeur had been there before, but their themes remained within an Arcadian realm of nature, exotica and childhood. Astuguevieille pushed the envelope with series such as the infamous Synthetics (Tar, Garage, Skai, Dry Cleaner and Soda), a blatant exhibition of the artificial nature of modern perfumery and an ode to the jarring yet oddly beautiful smells of urban life. If you think no one would wear the Synthetics, think again: within half an hour in the Parisian Comme des Garçons perfume shop, I saw three people walking away with them. Perhaps the most perverse thing about them is that they *do* develop, in the midst of those synthetic notes, very natural effects as well.

'Imagine being asked about your perfume at a social event. When you answer Garage, it's very provocative!' says Astuguevieille. 'The name acts as a wall label does for an artwork. It displaces things.'

He *does* consider the fifty-scent line-up he's produced over the past twenty years as he would an art collection: this play with the codes of the art world is, in fact, a deliberate strategy. When the first Comme des Garçons fragrance was launched in 1994, every bottle was shrink-wrapped differently and randomly, and most people kept them in their wrapping to exhibit them. 'I kept asking my friends what they thought of the smell, but none had torn the wrapping!' recalls Astuguevieille. Comme des Garçons Eau de Parfum went on to become so popular in the art world that, for a while, before I caught a whiff directly from the bottle, I was convinced that this was what galleries smelled like …

Comme des Garçons was not the only house to focus on 'found smells'. Across the Atlantic, the American Christopher Brosius conceived the very popular Demeter Fragrance Library, a resolutely figurative collection of captured odours like Dirt, Tomatoes, Baby Powder, Angel Food, Gin and Tonic, Play Doh or Hershey's. This figurative approach based on the pleasant surprise of instant recognition – 'Hey! That's exactly *it*! Wow!' – was an intriguing concept as well as a clever marketing move aimed at consumers who were put off by the abstraction of mainstream perfumery or its 'wear this and you'll pull' advertising. Tellingly, Christopher Brosius later launched a company called CB I Hate Perfume.

In their various ways, houses such as L'Artisan Parfumeur, Comme des Garçons, Demeter or CB I Hate Perfume play on the notion of perfume as Proustian experience – no longer a memory sprung from a chance encounter but a deliberate, (self-) conscious seeking out of it – as well as on a figurative stance that expands the repertoire of what can be represented in perfumery.

But the approach pioneered in the mid-70s by Jean Laporte has also generated a different type of figuration, driven by the exploration of specific notes. These scents zoom in on restricted zones on the scent-map, much in the way Dominique Ropion has done with Une Fleur de Cassie or Carnal Flower, to reveal the micro-cosmos hidden within. They remain figurative, in the sense that you can recognize what they're about. But because they explore olfactory archetypes in such depth, they verge on abstraction.

At least, that was my experience with Mathilde Laurent's Les Heures de Parfum ('The Hours of Perfume') for Cartier. I have a soft spot for Mathilde, not only because the maverick Corsican beauty with a Debbie Harry platinum streak in her hair is one of the most gifted perfumers of her generation, but also because she was the first to make me feel like a proper art critic. The first, in fact, to push me through the looking glass.

After my fateful radio show with Bertrand and Octavian, Mathilde wrote me an email to congratulate me for speaking out for the cause of creative perfumery. I wrote back to thank her for the marvellous Guet-Apens she'd composed when she was at Guerlain. She was touched. She didn't think anyone would remember a fragrance she'd done ten years previously, a limited edition she hadn't even been officially credited with at the time, though it later reappeared under the name Attrape-Coeur … attributed to Jean-Paul Guerlain (it has since been discontinued). Guet-Apens was everything I loved about Guerlain: Mitsouko's tender peach on a smooth-as-caramel amber shot with metallic glints of iris and violet. When she composed it, Mathilde was Jean-Paul Guerlain's assistant. She'd been hired after an unpaid internship before she'd even graduated – she was *that* good. She thus had the privilege of being trained by a man who had himself been taught by his grandfather Jacques Guerlain, a direct line of transmission from the pioneering days of the art. After leaving Guerlain in 2004, Mathilde was hired almost immediately by Cartier to compose bespoke perfumes. In 2009, the jeweller gave her carte blanche to create a line of thirteen scents, one for each hour of the day plus an imaginary one, Les Heures de Parfum.

As soon as I received samples of them, I realized that, despite having worn Guet-Apens for years, I did not know this woman. Who *was* Mathilde Laurent? How did her wonderfully quirky mind work? These were tough, cerebral, non-gendered, non-floral, unclassifiable: I was stumped. I could recognize the notes all right. That was the easy part. But as soon as I stuck my nose in them, they seemed to fly apart, each shard mirroring and distorting a fragment of the initial story. And then it dawned on me: I had experienced this very phenomenon with the late French Pop artist Alain Jacquet. In his best-known work, a variation on Manet's *Le déjeuner sur l'herbe*, Jacquet photographed the scene, re-enacted by real people, printed out the photograph like a movie poster then amplified the dotted pattern effect of the photomechanical print, thus revealing the irregular shapes of the dots. The perception of the picture depends on the distance at which you stand from it: zoom in and it dissolves in an abstract shimmer of irregular coloured dots. Zoom out and you perceive the image. There is a similar play on distance in L'Heure Brillante ('The Bright Hour'), for instance. At first sniff, a green citrus. But follow its evolution and it becomes a carousel of things that say 'green' and 'citrus' in every possible pitch: lime, petitgrain, verbena, citronella, cut grass ... Galbanum pops up, then green bean. Two bits of unripe fig are struggling to join up but don't because the milky note needed to make up a fig has been left out. The whole set is precision-assembled so that you never lose sight of the *idea* of 'green lemon-ness'.

But the scent that transfixed me was La Treizième Heure ('The Thirteenth Hour'). First sniff: lapsang souchong tea. Dive in: a smoky supernova expanding from a scorched leather core. Tar. Clove. Black tea. Juniper. Vanilla, which has a smoky facet, wrapped again in narcissus absolute, which picks up the vanilla's smokiness through paracresol, the 'horsy' single-malt whisky note. Vanilla and narcissus also connect through a common tobacco facet – another allusion to smoke. The blend is topped off with bergamot, which connects with the tea note for an Earl Grey effect. The idea behind La Treizième Heure is the 'fume' part of perfume: it reaches out into the distant past, from the smoke of burnt offerings all the way back to the first fire lit by prehistoric men. But it is *also* a tougher, edgier version of Shalimar with its bergamot, vanilla and leather structure, shorn of its powdery, balsamic opulence. After all, Mathilde knows its formula inside-out since not only has she weighed the materials for vats of the stuff in the Guerlain factory, but she's also composed an interpretation of it, Shalimar Eau Légère. Thus, La Treizième Heure is a *reflection* on the art of perfumery, on its history as well as on what it can say today. And this is what characterizes the perfumers who are moving the art forward: the fact that they are displaying, in a legible way, the process that drives their work.

None has gone further along this line than Jean-Claude Ellena, the in-house perfumer of Hermès and 'spiritual heir' of Edmond Roudnitska, whose streamlined approach he has carried on. Fittingly, Ellena is the author of the current edition of *Que Sais-je? Le Parfum*, which replaces Roudnitska's. In it, he explains the principles of his creative process – like contemporary artists, he has figured out that an oeuvre can gain in value when it comes with its own instruction manual …

What he calls his 'haikus' spring from an associative process reminiscent of Surrealist analogy: 'When I crumple geranium leaves between my fingers, I smell geranium, of course, but also black truffle, and truffle reminds me of the taste of olive oil; this, in turn, reminds me of castoreum, which has the smoky smell of birch, etc. The association of birch and geranium is an interesting accord.'

Ellena's style could be summed up in three words. Transparency: he dreams, ultimately, of reproducing the smell of water and his compositions have often been described as watercolours. Traceability: he discloses his sources of inspiration and some of his materials, including the synthetic ones. He has also been unusually forthcoming about displaying his creative process, allowing the writer Chandler Burr to follow the development of one of his fragrances for Hermès in *The Perfect Scent* and writing two books of his own, the pedagogical *Le Parfum* and the more personal *Journal of a Perfumer*.

The third operational concept in Ellena's style could be called minimalism. His aesthetics are those of the short-cut: he seeks to produce the maximum effect with the smallest amount of materials, and has famously reduced his palette to fewer than two hundred materials. But he objects to being called a minimalist, as he told me in an email after I had published an article on his work. Point taken: Minimal art, in the art-historical sense of the term, refers to nothing beyond the literal presence of the piece and does not evoke an outside referent, whereas Ellena's perfumes, however stylized and allusive, never sever their ties to representation. Let's call it concision, then.

If anyone can be called a Minimalist, it's Geza Schoen, a German perfumer who out-Ellena-ed Ellena by reducing the formula of his 2005 Molecule 01 for the brand Escentric Molecules to a single raw material. The velvety iso E super softens other materials, appeases tensions between musk, woody and floral notes, and makes perfume blends light as clouds. It is extensively used along with fresh green floral hedione and synthetic musks to create today's light, airy textures: the three materials act as 'fillers' and smooth out any bumps in the formula. (Scene from a lab: a perfumer chomping on his pencil. He still has to find four per cent of his formula. Hell, let's just stick in more iso E super). Geza Schoen's tough, authentically Minimalist gesture was to rip the utterly synthetic, discreet but ubiquitous iso E super out of its hiding place and display it for its own olfactory properties. Molecule 01 represents nothing, refers to nothing. In a riff on Frank Stella's words, Schoen could say, 'What you smell is what you smell.'

The second perfumer to explore the realm of the anti-perfume is, surprisingly (but only if you've never met her), Isabelle Doyen, who has been turning out the lovely and utterly *presentable* Annick Goutals for nearly two decades from the tiny, shambolic Parisian studio she shares with Goutal's daughter Camille, her business and creative partner at Aromatique Majeur. Isabelle, a slight, quiet woman who likes to listen to the French rap band

NTM while composing, wouldn't be out of place in the art crowd. In fact, when I *did* introduce her to an artist friend, she told me, 'Yup, she's one of us.'

Isabelle Doyen had a couple of ideas up the sleeve of her leather jacket that would indeed be more at home in an art gallery than on department-store counters. Fortunately, the tiny Swiss independent house Les Nez, owned by René Schifferle, gave her the opportunity to develop them. When Isabelle told René she'd always wanted to make a perfume that smelled of nothing, he gave her the go-ahead.

With L'Antimatière, Doyen explores the very limits of what can be called a perfume and still *act* as one though it is almost, if not entirely, imperceptible. While she won't reveal what's in it, she *does* say that it is composed of only five materials, all of them base notes: in other words, molecules so heavy they can take hours to fly off in the air and reach our noses. But this 'anti-matter' is present nevertheless, and will subtly skew any other scent that passes through its gravitational field while oddly amplifying one's own natural smell into the impression of a ghostly human presence …

Along with Geza Schoen's Molecules collection, L'Antimatière is one of the closest things to a piece of art the perfume industry has commercialized. And like any other piece of contemporary art, it sprung from its author's *need* to do it. 'It's there,' says Isabelle, 'take it or leave it. I don't have to convince, I don't have to justify. I made it because it was necessary.' To her, L'Antimatière is the olfactory equivalent of the Argentinian writer Jorge Luis Borges' aleph, a point in the world that contains the whole universe: 'A smell that contains every smell …'

Isabelle Doyen's obsession with Borges has also led her to conceive 'an outlaw perfume in progress' for the Turtle project, 'an open and chaotic network of diverse but interconnecting ideas, people, projects, events, and venues' woven both online and in the real world by the American filmmaker Michael H. Shamberg. The Turtle Vetiver series is based on another Borges story describing a walk on a never-ending street in an imaginary country. People go past buildings that seem to be of the same colour. But when they turn around, they realize that, in fact, the buildings are in a scale of colours so subtly different that they couldn't see them change as they were going by. Similarly, the Turtle series consists of subtle variations starting with the nearly raw vetiver oil, which Isabelle picked as a starting point 'because it is so complete that you can wear it as is … I don't know where I'll stop. I might end up with a totally different note!'

As she sits in her lab telling me about the people she's met thanks to Turtle Vetiver – when the eighty bottles of 'Exercise 1' sold out, she started blending fresh ones for the members of the Turtle network passing through Paris – I realize that the scent is generating what the art theoretician Nicolas Bourriaud has called relational aesthetics. In a nutshell, relational aesthetics concern artworks that subvert the classic one-to-one relationship between the piece and its viewer by integrating within their very conception the relationships they elicit *between* the viewers. An installation or a performance, for instance, does not exist autonomously: the movements, laughter, discomfort, exchanges, even the theories they inspire may all be part of the piece. In the same way, Turtle Vetiver, by drawing its aficionados into a loose, real-world network whose node is Isabelle's lab, goes beyond its nature as an adornment, a badge of olfactory identity, or even the matrix of a potentially unlimited series of variations on the formula. It has always-already been conceived to generate encounters, narratives. It creates community.

But then again, that's what perfume *does* – always *has* done, if you think of incense-burning. It's just that it took us a century to realize it. And to do something about it: that's when the online perfume community stepped in …

21

Why aren't there any critical reviews of perfume in the press, like there are for books, restaurants, wines or films? Let me put it this way. Once upon a time, a women's magazine I worked for had set out to test (unscientifically) everything from the pulling power of famous actors to the warmth of fake furs. In a bid to beat consumer report magazines at their own game, our publisher kept prodding us to find 'a face cream that gave women pimples'. We found a hapless journalist who'd caught a rash from a big-brand anti-wrinkle serum and published her story. The detergent-manufacturing behemoth that owned the brand immediately threatened to retrieve every single page of advertising for every one of its brands from every single magazine the press group owned around the world. Official apologies were issued. My chief editor got fired. The magazine haemorrhaged advertisers, and barely survived another two months. End of story.

There are no critical fragrance reviews in the press because the press depends on advertisers, and beauty features are essentially meant to keep advertisers happy. This isn't to say there aren't beauty editors who are knowledgeable about fragrance. And they do manage to slip in a few products from non-advertising brands that they actually love. That's as far as they can go. Any straying from florid praise might discombobulate the luxury brands that keep the glossies alive. On the other hand, I can no longer keep track of the number of perfume blogs and websites. It seems every time I peer at the screen, another one has popped up.

There was only a handful, three or four maybe, when I stumbled into them in mid-2005 after Googling Luca Turin, a biophysicist who'd published a perfume guide in French in 1992. Luca's reviews were irreverent, lyrical, side-splittingly funny: I'd never realized before reading them that it was possible to talk about perfume that way and, every year, I hoped he'd publish an updated edition. My search yielded Luca's blog, and when I'd finished reading it, I started clicking on the signatures of the people who left the most interesting comments. That's how I found Octavian Coifan, who went on to become a friend in the real world, then Bois de Jasmin, Now Smell This, Perfume-Smellin' Things, the Perfume Posse ... Within two years, my perfume collection had shot up to well over two hundred bottles. Perfume bloggers were the supreme enablers, tracking the obscure, the cultish, the vintage, cutting through the purple prose of press releases and magazine blurbs. They sniffed, they swooned, they dished the dirt. And as they went along, they invented a new language to talk about fragrance.

Most of the pioneers were members of Makeup Alley, a discussion board which features a reviews section. Now, reviewing cosmetics is a fairly straightforward business. Makeup and skincare make product claims: you're meant to see the result. And if a moisturizer turns your face into an oil slick, you can tell. Perfume, on the other hand, does nothing but smell, which is why its advertising relies exclusively on the three aspirational 'S's: stars, sex and seduction, with a side helping of dreams or exoticism. To speak about it, Makeup Alley's reviewers and their blogger offspring had to devise strategies that went beyond 'Would I buy/recommend it?' Descriptions, impressions, analogies, short stories, snippets of real-life testing, bits of history, parallels with music or literature … The styles could veer from 'Gal in the Street' to pure poetry. It was as though fragrance, because of its invisibility, mystery and evocative powers, had become a sort of *writing* generator that went far beyond its object. The very nature of the object seemed to attract a particularly literate community of amateurs, in the noblest sense of the term, 'one who loves'.

But perfume is devilishly hard to discuss. If you don't have your own mental catalogue of olfactory references, which is to say, smells linked with words, you won't get much past 'sweet', 'soft', 'screechy' or 'soapy'. One adjective comes from taste, the other from touch and the third from hearing; the fourth refers to a thing. There are very few words specifically related to olfaction in modern Western languages: descriptions draw from the vocabulary of other senses or from 'real-word' referents. Yet another problem is the names of notes that refer to nothing in common experience: amber, for instance. The word doesn't designate the fossilized resin of the same colour, but two different things. Ambergris forms in the digestive system of whales when they swallow something that irritates their stomach lining, like cuttlefish bones. They eventually expel it; it rolls around in the sea for years, bleached by sun and salt, and eventually washes up on a beach in the form of greyish brownish lumps. The lumps themselves are pungent things that smell, according to Jean-Paul Guerlain, of 'rye bread and horse manure'. Tinctured in alcohol, ambergris releases a delicate, warm, soft, slightly saline scent. For centuries, it was an essential material of perfumery and some houses still use it in their costliest blends, but few people have smelled it. Amber can also designate a blend of cistus labdanum (rock rose) resin and vanillin, invented by perfumers in the 19th century, a totally abstract smell that refers to nothing in nature. Yet 'amber' is a term commonly used to describe perfume notes and can refer to a slew of materials or accords. But go explain that to a novice who has smelled none of these.

Write about perfume and you'll be caught between your own limitations, those of your readers and the fact that, usually, they won't have the fragrance on hand to compare their impressions with yours. Even if you had the actual formula under your nose, you'd need training to decipher it; even if you had all the actual materials on hand as a reference, you'd still strain to figure out why the fragrance produces the effects it does. That said, perfumes aren't made for chemists, and if you write 'violet, iris, wood, leather' instead of 'methyl-ionone', you'll have a better chance of conveying the feel of a scent. Even with no knowledge whatsoever of raw materials, a writer with a keen olfactory memory, a good repertoire of fragrances and a way with words can write an evocative review. Connecting a scent with emotions, impressions, atmospheres ... isn't that why we wear it? Isn't it all *subjective*?

'If you smell it, it's there,' smiled the impeccably elegant Jacques Polge, Chanel's in-house perfumer. I'd just written a piece on the Exclusives, a high-end range sold only in the Chanel boutiques. What had intrigued Mr Polge was that it had been, unusually, published in a contemporary art magazine. When he called up the PR department to ask who I might be, they couldn't answer. But a friend of mine who worked at a rare-books shop where Polge had been a client for years spoke to him about me. That's how I ended up with an invitation for tea. Monsieur Polge would probably have been more inclined to discuss another one of our common interests, early-20th-century French poetry, but never having met a perfumer before – and the *Chanel* perfumer, no less – I endeavoured to draw as much information as I could, though not very successfully. 'If you smell it, it's there' was, frustratingly, all he conceded when I asked him whether there was such or such a material in one of his perfumes. I could have read his answer as a dismissal ('Little lady, just enjoy the stuff and leave the rest to the specialists'). Or it could've been I was right but that his policy was never to disclose that type of information. I chose instead to interpret it as an invitation to trust my nose.

But I don't buy into the 'it's all subjective' spiel. Sure, you can judge a painting based on whether you'd want it on your wall, a piece of clothing or a perfume based on whether you'd like to wear it. But just because you don't want it in your life doesn't make it bad. And it's not entirely impossible to consider perfumes beyond their 'like/don't like' Facebook status, which is to say, beyond their nature as consumer products.

When I first started covering fashion shows, I went at it like a girly-girl, picking the outfits I'd wear if I could afford them (consumer credit meant I could, and I'm still paying off that debt). But I co-wrote the reviews with a seasoned fashion journalist and she taught me how to ask other questions than 'Would I wear it?' The same questions you could ask of a perfume (or an exhibition, or a movie, for that matter): what intent does it set out to fulfil? How does it achieve its effects? How is it situated within the perfumer's body of work? How does it fit in with the history of the brand or its identity? How does it compare to the current season's offerings? Does it bring something new? What relationship does it bear to the history of the field? Can any light be shed on it through other creative fields?

These are objective questions and they can be answered objectively. We're not always able to, sometimes because we don't know enough, sometimes because the perfume doesn't ask them. Clearly, this approach won't answer queries like 'Does this stuff need to go on my most-wanted list?' But if at least a share of the art of perfumery is to be snatched away from the lacquered claws of the business-school Talibans it needs to be approached with something resembling art criticism: the more knowledge is collected, clarified and transmitted to the public, the more chances there are that at least part of that public won't accept unoriginal products, thus encouraging the industry to trust the perfumers at least some of the time.

When the whole blogging phenomenon started taking on proportions significant enough for the industry to notice, quite a few professionals were either dismissive or dismayed, and generally reluctant to acknowledge it, much less engage with it. Indie perfumers, on the other hand, were active participants in the scene from the outset for obvious reasons: they have no budget for PR and their products are mostly sold online.

The most striking case is Andy Tauer's. The self-taught Zurich perfumer started an online diary chronicling his creative process, dropped in to comment on other blogs, and sent their authors samples of his compositions. His talent, sweet disposition and openness endeared him to his public: unlike the stars working for big labs or Parisian niche houses, he was accessible through his blog and he was *their* discovery. Since then, Mr Tauer, a pure product of the online perfume culture, has been the poster boy of indie perfumery.

But the indie boom via online buzz is not the only way in which the thriving internet perfume culture has changed its object. In fact, something in the very nature of perfume may have shifted over the past few years.

Fragrance has always been a social medium. Though most people would say they wear it for themselves, it necessarily enters the social sphere since it is airborne, as the philosopher Immanuel Kant huffily remarked back in the 18th century: 'Others are forced, willy-nilly, to participate in this pleasure. And this is why, being in contradiction with freedom, olfaction is less social than taste, where among many dishes or bottles a guest can choose one that he likes without others being forced to share the pleasure of it.'

Yet apart from the occasional 'You smell lovely' or 'Yuck, is there something *dead* in here?' the social interactions produced by fragrances remain mostly unspoken. Except, that is, within the loosely knit network of tens of thousands of people who style themselves 'perfumistas' after the neologism 'fashionista', from the Spanish suffix for 'partisan of', as in Communist. (I'm not wild about the word and I loathe 'perfumisto': the suffix '-ista' applies to both genders.)

Fragrance generates community and this community is generating an ever-expanding volume of discourse. In fact, you could say the discourse on fragrance has been hijacked from its traditional owners, PR and marketing departments, beauty editors, trade journals ... Each review or comment transforms the perception of a fragrance, its description and the story that surrounds it. Whether it is the lovingly nurtured brainchild of a perfumer or the Frankenscent pieced together by a harried team on a budget, perfume is no longer just liquid in a fancy bottle. No longer just a product packaged and advertised by a brand. No longer just the stuff you spray on before a date. Each atomizer has become a kind of wormhole to a parallel world bursting with words, feelings, stories and people who may know each other in real life, but most likely not. In a way, you could say that the critical and social discourse that now surrounds perfume enables perfume to exist more fully today than it ever has.

What is, after all, a fragrance in a bottle? The brainchild of a perfumer, who's composed a song, a poem, a story out of smells. Thousands of people may buy it: at least, that's what he hopes. It may become a standard like Chanel N°5: he hopes for that too. But as long as it stays in that bottle, perfume is nothing, just as a song is nothing until it is sung and heard. It must be borne by skin, carried by air, perceived by noses and, most importantly, processed by the minds of those who breathe it in. The story told by the perfumer blends with the ones we tell ourselves about it; with our feelings, our moods, our references, our understanding of it. Once it is released from the bottle, it becomes a new entity, unique despite having been poured into thousands of bottles. We make it ours, like a singer sings a song: we are the *performers* of our perfumes.

22

'Skank: derogatory term for a (usually younger) female, implying trashiness or tackiness, lower-class status, poor hygiene, flakiness, and a scrawny, pockmarked sort of ugliness. May also imply promiscuity, but not necessarily' states the online Urban Dictionary.

But for perfume lovers, 'skank' has taken on another meaning. We're not talking white trash here: we're talking about what stinks. And we mean it in the nicest possible way.

Cumin: sweat. Jasmine: poop. Civet: ditto. Narcissus: horse dung. Mimosa: used nappies. Costus: dirty hair. Blackcurrant bud: cat pee. Honey: public urinals. Grapefruit: BO with a hint of rotten egg (it contains mercaptan, the sulphurous molecule used to scent the odourless natural gas so that we can detect a leak).

It was my friend March Dodge of the Perfume Posse who gave the word 'skank' its new meaning for perfume aficionados in a post about, of all things, some of the most widely revered masterpieces of perfumery, in which she detected something she termed 'the Guerlain Skank', 'a rump-grinding, head-shaking invitation to a booty call, no matter how politely the scent's been dressed up at the opening.'

No need to call in Dr Jellinek and his theory about erotic materials: perfume lovers, scrambling to catch up on all the classics and on the new, weirder niche stuff, had figured it out all by themselves. The love of skank is one of the most intriguing manifestations of the community's relational aesthetics dynamics. It isn't only a convenient term for 'somethin' dirty in mah perfume', as March says: it's a standard by which perfumes are judged, but also by which perfume lovers position themselves. And it's probably telling that the notion of skank originated in the hyper-hygienic USA.

In France, everyone's got at least one beloved family member who sported old-school scents, many of which featured the pungent animalic notes that have been edited out of contemporary commercial products: it is part of the Gallic cultural heritage. But when the American perfume *aficion* set out to explore the classics, especially in their vintage form, there was some dismay but also a challenge to prove one's mettle by actually *embracing* the skank. Perfumers often say there are no stinks, only odours to explore. Perfume lovers take the same attitude, along with the touch of snobbery every hipster cultivates – the general public may turn up their noses at the whiffy stuff, but *we* know better. In the words of one of the Marquis de Sade's libertines: 'We love what no one else loves, and this adds to our pleasure.'

This desire to push back one's limits in order to experience new forms of pleasure *is* reminiscent of certain sexual scenarios, but I believe there's something else at play in the impulse to sublimate our interest in smells into an aesthetic pursuit. Overcoming our aversion to stink through its incorporation into beautiful compositions could be a way of not renouncing our more primal instincts; of drawing the pleasures that Western civilization considers base, and that our education leads us to reject, into the life of the mind – hence the miles of words written on perfume by enthusiasts. Of course, not all of those words concern skank; very few, in fact. But our very obsession with scent does point towards a peculiar form of libidinal investment, which doesn't mean we derive sexual pleasure from our scented pursuits, but that perfumes engage some deep-seated vital energy, the libido. Skank is just the ultimate expression of that drive; its tell-tale symptom.

It also provides an excellent excuse for a well-educated group engaging in a refined, costly hobby to indulge in some cheerfully regressive pee-and-poop talk. Listen in to the chat. The most iconic perfumes in history are treated, literally, like crap. Roudnitska's leathery, cumin-laden Eau d'Hermès smells like 'Robert Mitchum's jockstrap in Grace Kelly's purse.' Guerlain's Jicky, 'like the cat crapped in a lavender patch'. Shalimar has been said to evoke nappies (used, N°2). Joy? The adult version of the excrement. Mitsouko exudes the sour smell of unwashed old ladies. Bandit is redolent of old ashtrays and soiled female undergarments. My mother's own Bal à Versailles also ranks high on skank: it has become the very epitome of 'unwashed panties' or, in the words of one Posse reader, 'cat butt morphing into cured horse manure'. To which March the Skank Queen replies: 'Maria is a true perfumatrix. She smells something that goes from cat butt to cured horse manure, and does she burst into tears? Run away screaming? Saw her arm off? Nope. She *takes notes and waits for the drydown*. My tiara's off to you, Maria.'

As for myself, I'm not averse to a bit of skank, but then, I'm the type who sniffs her lovers greedily and turns around to take in the wake of attractive strangers. A little blast of *eau d'humanité* never hurt. I'm more put off by fragrances bleached of anything that could smell remotely gamey, the anorexic juices that feel like they've gurgled fabric softener.

My own entry in the unwashed panties category is the vintage version of Elsa Schiaparelli's 1937 Shocking. 'That Italian artist who makes dresses', as Gabrielle Chanel sneeringly called her, introduced Surrealism into couture by collaborating with Salvador Dalí on some of her more outlandish models (lobster-adorned jackets, a shoe worn as a hat, another hat resembling a lamb cutlet, complete with frill). The hot-pink box that shrouded Shocking expressed the indecent intensity of desire sung by her Surrealist friends. The scent itself was as flamboyantly immodest as its namesake pink. The perfume bottle expert Jean-Marie Martin-Hattemberg (quoted by Richard Stamelman in *Perfume: Joy, Obsession, Scandal, Sin*) calls it 'the first sex perfume'. And it is. Imagine a Parisian woman who has just spent a night in the arms of her lover. It is too late in the morning for her to go home and change, too late even to take a shower. She hurriedly splashes on a rose and lily-of-the-valley fragrance, even dabs a touch of it on her silk briefs. When she comes home and slips out of her silken lingerie ... She smells of Shocking.

Shocking does reek of *gousset*, the small triangle of fabric sewn into the *petite culotte*. Its combination of rose, ambergris, honey, civet, musk and sandalwood is possibly the closest evocation of the female bouquet ever devised by classic perfumery, barely veiled by the green floral fig-leaf of a lily-of-the-valley and gardenia heart. One can only imagine the effluvia wafting up from Schiaparelli's place Vendôme salon as the artist Christian 'Bébé' Bérard 'put scent on his beard until it trickled onto his torn shirt and the little dog in his arms' or 'Marie-Louise Bousquet, the witty hostess of one of the last Paris drawing-rooms, [pulled] up her skirts and drenched her petticoats with it,' as the designer recounts in her autobiography, aptly entitled *Shocking Life*.

Mainstream perfumery has veered as far away as possible from skank, with a few noted exceptions (the late Alexander McQueen's now discontinued, cumin-laden Kingdom was a flop). Niche perfumers, however, have often explored the territory with such animalic blends as the infamous Muscs Koublaï Khan ('unwashed Mongol warrior after six months on the saddle') or L'Artisan Parfumeur's Al Oudh ('camel driver's armpit'). But the Swiss indie perfumer Vero Kern definitely raised the bar with Onda, possibly one of the most challenging – and ultimately rewarding – compositions in that particular genre.

Vero, a warm, rangy woman with expressive features and an exuberant laugh, came to perfumery after having worked for Swissair, converted into massage therapy and veered off into aromachology. She found out she loved the essences she worked with for their beauty even more than for their therapeutic properties, took a course at Cinquième Sens and launched her tiny house with Kiki, Rubj (pronounced 'ruby') and Onda. Vero's perfumes have soul and, like souls, they're full of sublime beauty and dirty secrets. Her Onda is about earth, flowers and flesh smeared in spicy honey. The honey and musk wrap the earthy notes of iris, patchouli, oak moss and vetiver in a human funk that makes you feel you've sunk your nose in the lustily worn and discarded garments of your lover – there *is* more than a hint of the *petite culotte* in there …

Onda's earthiness, in both the literal and figurative sense, points towards another area of skank: the zone on the olfactory map where overripe fruit, rotting flowers, decaying vegetation, mouldy earth and stagnant water conjure the miasma of similarly corrupted animal flesh. This reminder of the destiny of all living things – a *memento mori* like the skulls Flemish painters placed in their vanities – stirs more anxiety than the odd hint of whiffy briefs, uncouth armpits or lavatories. Perhaps because its vegetal origin makes it more inhuman: Nature swallowing us whole.

'I put it on, OK, good enough, big swampy flower, and then … the decay goes … deeper? Sharper? Wiggly? Something happens to make the swamp sweeter and more smothering in a way that I find vaguely panic-inducing,' commented a Posse reader about a fragrance that has gained cult status, Sandrine Videault's Manoumalia for Les Nez.

Sandrine Videault, who was taught by the great Edmond Roudnitska, has always straddled the frontier between art and perfumery. She has done very few commercial fragrances, but has collaborated with artists like Fabrice Hybert and Hervé di Rosa in olfactory installations and authored several of her own, most notably at the 2000 FIAC (the French International Contemporary Art Fair); she is also an olfactory archaeologist of sorts, who re-created the ancient Egyptian kyphi for the Cairo Museum. She also stands apart from the industry geographically since she has settled in her native New Caledonia, a French territory in the Pacific. This was the springboard for the first ethnographic perfume, based on the rituals of the islands of Wallis and Futuna, in a move reminiscent of the Cubists when they reached for African art as the means to break down the codes of figuration. Built around a reconstitution of the fagraea flower which grows on the 'Taboo tree', Manoumalia reprises the notes of Wallisian rituals: fagraea, sandalwood powder, vetiver, and an old French perfume Wallisians are so fond of they often wash their hands in it, L.T. Piver's Pompeia. The result is both suave and strange: a creamy tuberose-frangipani-gardenia accord ripe with mushroom and gasoline notes on a bitter, leathery vetiver base and a buttery trail of sandalwood; the offspring of Bandit and Fracas gone native.

Manoumalia is a sophisticated composition yet it is also so primal that it elicits amazingly violent reactions. Rubber, drain cleaner, faecal matter, rancid butter, cheese, mothballs, formaldehyde, urinal cakes, ashtray, rotting animals and even the 'sweetish, faintly bloody and meaty' smell of afterbirth ... There isn't an evil stench Manoumalia hasn't been compared to. But as an online commenter points out, these amazingly negative reactions are actually a testimony to its stunning realism: 'I am always struck by the rotting, faecal, vegetal, death/birth/death/birth smell of the tropics. I think it's because they are touted in the media as being sweet and charming/flowery when in fact they are savage and terrifying in their desire to regenerate.'

'Manoumalia stirs up passions and that seems positive to me,' Sandrine told me in an email after reading these comments. 'All the associations are either accurate or justified ... Sleeping with tuberoses or fagraeas by your bed can be unbearable because they are so powerful and can unfold faecal or rotten facets. Those unpleasant facets are not olfactory hallucinations ... The point is to get back to the form of the fragrance rather than staying stuck on facets. All of this depends on our state of mind, or even the state our soul is in at the moment of olfaction.'

And this may well be what lies at the core of the powerful feeling of repulsion the insanely beautiful Manoumalia induces in some wearers: the obscenity of flowers *exposed* as perhaps never before; a trail of damp, red-in-tooth-and-claw tropical nature that could send you off muttering, 'The horror, the horror', *Apocalypse Now*-style ... What Manoumalia conjures is an olfactory archetype, and one that speaks deeply to us. But the emotional tenor of those reactions is cultural: what triggers anxiety in Americans causes Wallis Islanders to break out in a huge grin; it can make a sixty-year-old biochemist from Southern India gush with tears in his eyes that he's been looking to capture that smell all his life ... It is precisely because Manoumalia was born of a quest to renew the vocabulary of Western perfumery that it includes odours that would have been sandblasted out of a more commercial product.

The need to avoid overtly unpleasant notes marks the limits of perfume as a contemporary art form. There is, however, one notable exception: Sécrétions Magnifiques, conceived by the maverick niche house État Libre d'Orange for its shock value. The scent is based on the smells of blood, saliva, semen, sweat and maternal milk. They are designated as such in the notes list, with an ejaculating penis as a visual to drive the point home.

Playful provocation is part of État Libre d'Orange's DNA. Their products bear names like Putain des Palaces ('Fancy hotel whore', a tribute to a song by Serge Gainsbourg), Charogne ('Carrion') or Don't Get Me Wrong Baby, I Don't Swallow. Some of their visuals are correspondingly graphic. Both overtly display what the perfume industry has been selling itself on for decades: sex. But they throw something into the mix that the industry has never allowed itself to draw on: an iconoclastic sense of humour. By putting together irreverently clashing concepts such as jasmine and cigarettes or incense and bubblegum, the scents themselves often play on the Surrealistic process inspired by a quote from the French poet Lautréamont: 'the chance encounter of an umbrella and a sewing machine on an operating table'.

But none are as deliberately shocking as Sécrétions Magnifiques, which has achieved such iconic status among perfume connoisseurs that it is the benchmark against which everything gagworthy is gauged, including by people who haven't actually smelled it. The fragrance blogger Katie Puckrick has even put out a YouTube video where she applies it live as if it were a stunt – the perfume-world equivalent of the MTV reality show *Jackass*.

But why do skank aficionados who pride themselves on having overcome human-cumin-phobia and gloat at indole over-doses reach for sandpaper whenever a molecule of Sécrétions Magnifiques brushes their skin? All the notes it contains are notes *they* contain – many of which they've actually swallowed. True, Sécrétions Magnifiques *is* disconcerting with its odd metal-lic and iodic notes, but not quite as literal as the visual implies. Could it be that identifying its notes as blood, sperm, maternal milk and so forth, rather than saying 'metallic', 'marine' or 'creamy', is what triggers such exaggerated reactions? In the London niche perfumery shop Les Senteurs, the fragrance adviser dons a latex glove to spray a blotter, as though the scent literally contained the bodily fluids listed and therefore presented a medical hazard.

Sécrétions Magnifiques breaks the boundaries between the liquids we squirt on and the ones we squirt out, but its author Antoine Lie never envisioned it as a literal rendition of the smell of semen. In fact, that was a point on which he disagreed with the owner of État Libre d'Orange, Étienne de Swardt. Lie says he was 'much more interested in what was happening inside: on the story of the internal fluids that provoke desire'. He therefore structured Sécrétions Magnifiques around a fictitious 'adrenaline accord' with saline, mineral aspects, 'which acts as a conductor body where all the other substances can soak'.

Sécrétions Magnifiques had been conceived as a buzz-generating oddity; both Lie and de Swardt thought it might only please five people in the world. It went on to become one of the brand's best-sellers, though some might buy it as a novelty item (the État Libre d'Orange flagship store sits at the edge of the Parisian gay quarter, le Marais). Still, I *have* seen some people come to the shop to replenish their stock. But Lie is well aware that the buzz is closer to lynching than love-in: 'I even read that someone like me should be locked up in an asylum ...' he told me. 'People say it's disgusting but, for me, the mechanics of internal fluids represent beauty in its purest state. Because in fact, *that's* what's true. When you feel an emotion, it's triggered inside, hormones circulate, blood pulses, you sweat, you get goose bumps ... That's what I wanted to express: that what happens inside smells like *that*. It's not disgusting. It *seems* disgusting to you, but it's something true: you don't cheat.'

Who says fragrance could not be as disturbing as any other art form?

23

If anything shocks me, it isn't a perfume that smells of spunk or funk. It is a perfume that's been face-lifted, liposuctioned and laser-resurfaced out of any likeness to its former self: vandalized.

Today, I went a stealth mission in the Champs-Élysées Sephora with a two-millilitre phial nestled in the palm of my hand. I needed a few drops of a 70s classic for my course at the London College of Fashion: my twenty-year-old students have never known the disco era. Most of their *parents* were barely pubescent at the time. So I made a grab for the tester bottle, took a sniff and let out a little yelp – so much for stealth. Surely this was one of the limited-edition variations the brand put out each year. I double-checked. The bottle had been changed, but it was definitely the original version in eau de toilette. Well, the juice had probably been cooked by the display lights. I grabbed the eau de parfum tester ... God knows that particular smell is seared into my memory. There are probably still molecules of it lodged in my bone marrow, left over from those three years working in a high-end department store at Christmas where I inhaled the stuff for days on end … It seems yet another classic has been spayed. I'll miss loathing it.

Your favourite perfume doesn't smell the way it used to? Don't try complaining to the sales assistants: they'll swear blind your 'chemistry' or sense of smell have changed and you'll go home wondering whether you've been hit by early menopause. But don't book an appointment with your gynaecologist yet: most likely, it's what's in the bottle that's undergone the Change. You just haven't been told about it.

Reformulation is the perfume industry's best-kept secret. You'd think altering the nature of a product without notifying consumers would amount to fraud. But since the people who have the competence actually to demonstrate the change work for the perfume industry, all we've got to go on is our nose. The practice isn't new: perfumes have always been reformulated. Labs tweak their products over the years. As companies change hands, formulas may be lost altogether. But, mostly, perfumes are reformulated either to scratch the bean-counters' itch or to prevent you from getting one. After all, it is a well-documented fact that entire populations were wiped out by pruritus after having dabbed on a drop of Joy.

The perfume industry had been merrily blending away when it suddenly decided to clean up its act by finding out which materials could potentially be harmful to consumers. Rather than have public authorities poke their noses into their formulas, it decided to self-regulate. The International Fragrance Association (IFRA) was founded in 1961: its members agree to comply with its standards on raw materials (i.e. which percentage can be used in the formula, according, for instance, to whether the product is 'leave-on' like fragrance and moisturizer, or 'rinse-off' like shampoo or soap). Based on data collected by the Research Institute on Fragrance Materials (RIFM) and its members, IFRA publishes yearly amendments to those standards. And each year, another material that's been used for decades, centuries or even millennia is restricted or banned. IFRA standards are not legally binding since it is a private industry organization, but several countries base their legislation on them.

Is there any way out of those standards? Well, unless you're an indie perfumer who makes up all her blends herself and sells them directly to her customers, if you have to produce a certain volume you'll have to get your oils made up by a lab, and that lab will almost certainly be IFRA-compliant. Even if you wanted to go ahead and use banned or restricted materials, you might not be able to find them, or have to pay through the nose for them, as producers stop growing, distilling or synthesizing them because they can find no outlets.

Neither allergists nor regulators and legislators care about the fact that putting the hex on hydroxycitronellal or oak moss will disfigure a slew of masterpieces. They're not even aware there's a problem. To them, fragrance is a consumer product and if an ingredient can give rashes, even to less than 0.1% of the population, they figure you can just replace it with a perfectly innocuous substance. The problem is that there are no perfectly innocuous substances and that, as you replace one thing with another, that new thing will trigger allergies in turn as more people are exposed to it. And you don't tinker with a perfume formula as you would with a detergent's or a weedkiller's. While the two latter might be as effective with a different ingredient, it's the former's very nature that is altered. Perfume formulas are delicate balances: change one material and the whole structure is skewed. And lest you think that only evil, lab-concocted chemicals are the culprits, think again: we're talking about bergamot, rose, jasmine, basil, tarragon, cinnamon, clove … vanilla! Though when IFRA tried to restrict *that* particularly popular material, or rather vanillin, the industry balked and the regulators backed off. It would have probably meant reformulating more than half the perfumes on the market.

The issue boils down to this: our society has given in to the zero-risk mentality. To decision-makers, whether corporate or institutional, public safety – and avoiding lawsuits – will always trump aesthetic achievement or cultural heritage. As the perfumer of a legendary house told me, 'It's no use trying to convince politicians to ease up on regulations. They'll reply that, since our products are useless, the least they can do is to be totally innocuous.'

Ah, yes. The perfumers. You'd think they'd be the first to complain that their palette is being reduced, that they're forced to reformulate their own products or do hatchet jobs on those of their predecessors. Do they speak up? Let's say there's a lot of grumbling and hand-wringing going on in the wings and a few polite throat-clearings in the press. Mostly, they get on with their jobs since they can manage to do beautiful things with what's left. It's not like they're going to take up pottery instead, is it? But they are starting to realize that certain styles will be rendered impossible: how can you play on the overdose of a material – and many masterpieces *were* the result of an overdosed material – when only piddling quantities are allowed? When the standards can change from one year to another? The younger perfumers have never used the restricted materials so they don't know what they're missing. And the older ones who *could* have spoken out while there was still time say that when the whole regulatory kerfuffle cropped up they thought it would blow over. It didn't. Regulatory organizations tend to take on a life of their own and to metastasize until they choke the body that hosts them. Besides, bringing up the issue of allergens and toxicity in the media, even to defend the innocuousness of perfumes, would only fire up the public's chemo-phobia. Never mind that the restricted allergens are often molecules present in natural materials, and that you can probably absorb more methyl chavicol in one bowl of pasta al pesto than by bathing in your old Eau Sauvage for a year. And never mind that peanuts, which can literally kill, are nowhere near being banned; that trees aren't being uprooted because their pollen gives hay fever to a significant part of the population. You've got to eat, and uprooting trees would bring every environmental organization out of the woods. But perfume is useless: no one's going to go up in arms for it. Who cares as long as the stuff still smells nice? Who cares that beauty is being blasted into oblivion by bureaucrats bent on protecting us from ourselves, and by luxury groups who make their money from new launches and don't care much what becomes of their classics? The brainchildren of perfumers are being hung, drawn and quartered. And with them, the olfactory heritage of generations, our memories, our emotions; for those of us who've worn a scent all our lives, it is our very aura that's being ripped out of our flesh.

The *omertà* on reformulations was broken by a man I'll love forever for writing that he'd 'risk scrofula' to get his old Brut back, though I have no love for Brut. Luca Turin provided perfume aficionados with their first cause célèbre, the fate of Mitsouko, by alerting them to its impending reformulation by Guerlain due to new restrictions on oak moss, a material essential to its formula. Mitsouko became the poster child of everything that had been good about perfumery and everything that was going wrong. It took on the status of the Ultimate Cult Perfume, the one fragrance lovers desperately struggled to 'get', as though not 'getting' it somehow disqualified them.

It is sadly ironic that the perfume community started coalescing at the time when the classics were being irreparably defaced due to a surge in regulations. We'd found the way to Shangri-la just as promoters were tearing it down. For a few months in the fall of 2005, I wondered whether I should stock up on Mitsouko before the old formula was pulled off the shelves. I was desperately broke at the time and had to skip several lunches to snatch a bottle before it underwent its surgical reconstruction. In all fairness to Guerlain, it did limit the damage by entrusting the reformulation to the senior perfumer Édouard Fléchier, and is now campaigning to get its classics exempted from regulations as part of the French cultural heritage so that their original formulas can be restored.

But back in 2005, *l'affaire Mitsouko* triggered a frantic vintage-collecting phase. For several months, I tried to get hold of as many old bottles as I could afford. I received parcels so regularly I ended up having a short affair with the cute mailman, who'd look at me, barefoot and dishevelled in my tatty vintage silk kimono, as though I were fully made-up in a satin marabou-trimmed negligee, and murmur, 'I *love* waking you up …'

If a bottle is kept away from the light and heat, perfume can actually keep for decades, though a few of the top notes may have turned. Hold out until they've evaporated, usually within a few minutes, and you'll discover the perfume as it was intended to be, with richer, more vibrant materials, a higher proportion of natural essences and some synthetics that are no longer used: nitro-musks, for instance, which literally made the other notes pop out in 3-D. Today, half my refrigerator has become a branch of the Osmothèque, the Versailles institution where old formulas are preserved and reproduced. But instead of holding fresh reconstitution, my collection is the final resting place of bottles produced in bygone decades: a cemetery of flowers.

There is something poignant about those rows of bottles, some of which are older than me. I bought most of them boxed and sealed: why did they go unworn and unloved? Take this bottle of Scandal wrapped in logoed Lanvin paper. Was it a disdained gift? Since the owner never opened it, she didn't even know the box held one of the most beautiful leather fragrances in history. Maybe it was never even given to her. Maybe she passed away before wearing it … More poignant still is the awareness that, every time I dip a blotter into one of those bottles, every time I extract a few drops with a disposable pipette to dab on my skin or show to my students, I am annihilating a bit of the past. Once the bottle is used up, there will never be another one. I can always engage in a bidding war on online auction sites with fellow perfume lovers or our arch-foes, the bottle collectors, Huns who'll just let the stuff *bake* in well-lit display cabinets. But there are just so many knocking about, all fading away, oxidizing, surrendering molecule after molecule to the air despite being sealed – what cognac distillers call 'the share of the angels'. Once they've gone, that's it: perfume is a lesson in letting go. Though, given the size of my stash, I could probably live to be one hundred and still be embalmed in the stuff.

But there is one bottle I'll cry when I empty. A bottle so rare that the mere fact I own it hatches plots to burglarize my flat. The French couturier Jacques Fath launched Iris Gris in 1947, but he died prematurely in 1954 and Iris Gris was soon discontinued because it was expensive to produce and a commercial flop, hence its extreme rarity. I had no hope of smelling it outside the Osmothèque until I happened on an open-air flea market under the aerial metro that runs past the Eiffel Tower. Eyes peeled, I wandered from stall to stall. My knees turned wobbly when I spotted it. No more than one fifth evaporated, sealed, boxed. I stole away with my precious. I knew I'd got hold of a myth.

I felt the unsealing of Iris Gris needed a witness and had invited Octavian Coifan. We met under the gilded mirrors of the café Le Nemours by the Palais-Royal. That first dip of the blotter knotted my stomach … Had it gone bad? I took one sniff and grinned like an idiot. Octavian practically started purring. It was impeccable. What first jumped out of the strip was the peach, as smooth as a Renoir model's downy cheek. Octavian, who'd come equipped, handed me blotters of orris absolute, irone (the molecule that develops in orris butter when it ages) and ionones (the violet smell) for comparison. And, magically, iris came to the fore, its slight metallic tinge softened by the peach. Musky, with raspberry, apricot and leather overtones, the merest touch of a floral heart and a tiny celery note … The effect was amazingly modern and spare: Iris Gris could've been composed yesterday. In fact, it could walk circles around any iris scent on the market.

Since that day, 'I'll let you sniff my Iris Gris' is the louche-sounding proposition that'll draw any perfume lover into my lair. I'm quite a tease about it and I've kept some people dangling for months before unscrewing that crystal stopper. Bertrand Duchaufour got a few drops because he'd told me he had the formula tucked away and I was angling to get him to make up a fresh bottle for me before he agreed to compose Duende. After that, asking him to mix me a fresh batch of Iris Gris would have seemed too greedy. I asked him anyway. And he *did* tell me I was too greedy.

Iris Gris was the fragrance I chose to face the greatest concentration of perfumers I've ever been confronted with at the French Fragrance Foundation gala. My scented wake had to intrigue the pros. The prospect of the best noses in the world diving towards my cleavage or nuzzling my neck was entertaining. So I decided to lavish a whopping 1.5 ml of my precious on my skin and hair: to wear it as though we were in 1947 and there were still buckets of the stuff up for sale. After all, the great Vincent Roubert hadn't come up with this velvet-skinned marvel so that it could finish its earthly days in my refrigerator …

Let's just say that it's a miracle I came back home smelling as divine as I'd walked out: I'd have thought every molecule had been snorted off me.

24

'*Ça manque de cul.*'

Cul is an exquisitely expressive French word – after the palate-slapping 'c', uttering the 'u' is like puckering up for a kiss, while the 'l' remains silent – that can mean either 'ass' or 'sex'. Monsieur pronounces it greedily. I doubt anyone knows the way I smell better than him: over the course of our long affair, he rooted round my body like a truffle pig. He is the first man I've ever tested Duende on, and he feels just as I do: 'not enough sex'.

It took me over two years to reach out to Monsieur again after I cancelled our last date. When I did, I sprayed the inside of an envelope with Carnal Flower, the last scent he'd given to me, and sent it to him. Within twenty-four hours, he'd answered the call of the tuberose. I joined him for dinner at Lapérouse on the Quai des Grands Augustins, the last Parisian holdover from the era when gentlemen entertained courtesans in private rooms. The scratched mirrors have never been replaced: this is how tarts checked to see if the diamonds they'd been offered were the real thing.

I'd sent him an olfactory message in that envelope. He greeted me with another. 'You're not wearing Mouchoir de Monsieur!'

was the first thing I told him after he'd pulled me onto his lap on the banquette and we'd laughed at the joy of touching each other again, even before the first glass of champagne. I wondered whether another woman had chosen it for him. Ironically, it was from the same Italian house as the fragrance that had suddenly appeared on the bathroom shelf during the uneasy period when my husband and I shared our flat while getting a divorce. Though at the time I had been having an affair with Monsieur for over a year, the invasion of my bathroom by another woman's gift had roused my territorial instincts: I'd symbolically pissed in the bottle with a squirt of clove-laden Coup de Fouet, which the Tomcat had always loathed. If he'd noticed, he'd never said.

And now, a year and a half after having resumed our acquaintance, Monsieur and I are sitting at the terrace of a restaurant near the Bastille in the early September sun. He's saying he's already smelled Duende on me. He couldn't have, but perhaps he's detecting something of Bertrand's signature: I've worn Nuit de Tubéreuse on occasion to meet Monsieur. Though it is purely olfactory, there is a faint, exquisite whiff of symbolic betrayal in carrying the scent of one man on my skin to entice another. This is a jungle thing, a marking of territory I only became aware of as I held out my wrist to Monsieur. And Monsieur is a subtle enough reader of my thoughts to have guessed at this subconscious double-timing even before I did. I've been ... sprayed by someone else. Nevertheless, the man who knows me best thinks Duende doesn't smell enough of *me*. Or of *cul*. But serendipitously, the mailman has just brought me a parcel containing a bit of the beast, which I intend to take to my next session with Bertrand.

The minute I ripped open the envelope last week, my Siamese cat materialized in the living room, howling. As I unravelled the bubble wrap around what looked like a tiny brown crumbly stone, she jumped onto the desk and tried to prise it away from me, purring, nipping and batting like a mad thing.

This was a gift from Dominique Dubrana, also known as Abdes Salam Attar, who practises the ancient art of blending natural essences in a small town above Rimini, in Italy. He is a convert to Sufism, the mystical branch of Islam; his compositions are profoundly meditative, moving back centuries into the sacred origins of perfumery. I've never met Abdes Salam, but we've exchanged emails. And now this gift: a non-identified, mystery material in a parcel. I had to lock up the cat before having a proper sniff, which yielded fascinating observations: facets of blood, acetic acid (present in vinegar), apple, flint, patchouli, vetiver, pepper, cow and horse dung, herbaceous notes, blackberry, blackcurrant bud, ink, leather ... I shot off an email to Abdes Salam to ask him what the stuff was. He replied he'd deliberately not identified it so that I would ask him the question. But by the time he answered I'd sussed it out on my own: the little rock now resting in a feline-proof box was African Stone, the fossilized urine of a rabbit-sized South African beast called the hyrax, or dassie. Hyrax colonies urinate at the same place, often for centuries, and their urine is jelly-like rather than liquid. Over time, it fossilizes and can be collected in chunks. It has been used in vernacular medicine for centuries and the perfume industry is starting to use it as a replacement for materials of animal origin, since it is cruelty-free. After all, it's basically recycled piss.

Ick? Hardly. I couldn't get enough of the warm, funky, somehow *feminine* smell of African Stone. I wrote back to Abdes Salam, whom I could imagine chuckling in his beard as he typed back that I must be a member of the olfactory perverts club ...

'Or maybe you're a pervert, period?'

I've just handed Bertrand the African Stone over our table at the upscale soup and sandwich shop next door to his lab. Are the pheromones in this stuff acting on his brain as they did on the cat's? I sometimes forget that he's, well, a *guy*, and that there are some words, like 'pervert', you can't utter in front of a straight guy without his cracking some kind of joke. He's got an odd glint in his eye. Thrust a new aromatic material under a perfumer's nose and he'll go into overdrive.

It's the first time Bertrand and I have managed to get together since the summer holidays and that fateful question he asked me: 'Do you know what you want?' Although I did write to him during the summer that I'd thought about it, had a clearer idea, and asked him to wait before working on new mods, he's gone ahead on his own.

We settle down into what has by now become a well-rehearsed routine. This time, there are four phials, numbered 11 to 14. The scent wafting up from the strips has become familiar as well, though each time I encounter a new version something has shifted. Today, a new layer of expression has been added: grapefruit to make the zesty top notes bitterer and a dry, crackling fizziness produced by a material called magnolan to magnify the floral note and intensify the incense. Each mod contains double the amount of magnolan put into the preceding one: three per cent in 11, six per cent in 12, and so forth.

'Just small tweaks. That's what you'll be seeing more and more of from now on, unless we want to change our course drastically at some point.'

I frown over my blotters. If this were something I was discovering in a shop, would my credit card levitate from my purse?

Bertrand dips a strip into a phial of magnolan. It smells both of wet stone, flowers and grapefruit, with a crackling quality, as though it could punch little holes into a blend to fizz it up. Interesting, but I still think Duende isn't sexual enough, I tell Bertrand, quoting my recent conversation with Monsieur. He shoots a curious glance at me.

'Who's Monsieur?'

I've never told Bertrand about him. We don't tend to discuss the details of our personal lives and, besides, I wouldn't quite know how to describe my relationship with Monsieur. So I squirm a bit, grope around for a pithy answer and end up twirling a raised hand in the air with a little shrug and what I hope is an enigmatic smile. For a second Bertrand looks at me expectantly and I brace myself for the wisecrack I swear is bound to come out, but he just shakes his head with a little smile in an 'Oh ... *women*' kind of way and gets back to the business at hand.

'For the moment, this must be as floral as possible – besides the fact that it can be very *cuisse de bergère* ['shepherdess's thigh'] or anything you please!'

'You mean *cuisse de nymphe émue*!'

The term means 'thigh of excited nymph': it designates both a variety of rose and a colour (which translates rather more prosaically in English as hot pink). Clearly, Bertrand has never heard it: he repeats it quizzically and bursts out laughing.

'Then we'll be doing *cuisse de nymphe émue*, no problem but, for now, I want to make the floral accord impeccable. Magnificent. It's got to be *unique*, unlike anything I've done before. I want you to feel as though you were walking down an alley of orange trees in blossom, with jasmine dripping all over the place!'

Quite. But what I'm getting is mostly banana. The effect comes from the jasmine and ylang-ylang. Bertrand has used a new quality of the latter, sourced through a fair-trade organization he's been working with in Madagascar. Though I think they're a little off-course in the story, I love those over-ripe fruit notes: they have an almost animalic effect.

'Yes, they can even be a bit *petite fille qui se néglige* [little girl who neglects herself],' Bertrand agrees. 'We'll play on that. But, for the moment, we're not there yet. Now that I've perfected the fruity note that covers everything up, we might add more incense – I started out with ten per cent and I'm down to one!'

As I wave the blotters under my nose, I catch a whiff of the African Stone that's rubbed off on my fingertips. It blends in quite nicely. I like the idea that it landed on my *cuisse de nymphe émue* just at the time when I was concerned about the perfume not being erotic enough, as though the perfume itself had called for it …

'Correct me if I'm wrong, but couldn't African Stone go into this?'

He thinks it over for a couple of seconds.

'Yes! Of course! But we'd have to know if we can use it. I always get slapped on the fingers because my products are too expensive.'

Price isn't the only factor. Natural raw materials have to be available in sufficient quantities and in a consistent enough quality to ensure the same formula can be produced over the years. More importantly, they need to have passed a battery of scientific analyses to demonstrate they aren't harmful. Actually, no animal substance in perfumery is – but you've still got to be able to prove it. Not to mention that, though African Stone is cruelty-free, some consumers might balk at the idea of putting an animal substance on their skin. Especially fossilized pee. Still, we agree I'll ask Abdes Salam about it and we go back to comparing the variations.

Bertrand dabs on numbers 11, 13 and 14, waves his arms around to accelerate the evaporation of the alcohol and holds them out to me after smelling each dab. Professionals don't take the dainty sniffs fellow perfume lovers indulge in during joint expeditions. They go for full-contact, snout-to-skin molecule-snorting. So I do it like I've seen it done: I grab Bertrand's left wrist, tug it towards me and run my nose along his inner arm. Though the contact is meant to be purely impersonal, it is disconcertingly intimate – a prolonged, inquisitive animal snuffling of the type you only indulge in with your lover or your child in the real world – and it hasn't become routine enough for me to feel like business as usual. This is the first time I get a whiff of Bertrand's skin: when we peck each other on the cheeks French-style, the gesture is warm and affectionate but subtly timed and choreographed to remain within the appropriate social codes. No time for sniffing out the beast.

As I flit from mod to mod, I tug one of Bertrand's arms towards me then the other in a strange little dance. When the analytical part of my brain finally kicks in, I find that N°14, the one with the highest dose of magnolan, seems to collapse under its own weight, sending all the facets flying in separate directions. And we're both thinking the banana note is a little over the top. This isn't Carmen: it's Carmen Miranda. Bertrand concludes he'll reinforce the animal, narcotic notes as I asked him to, tune down the banana and raise the percentage of incense.

'Take these,' he says, handing me the four phials, 'wear them, follow them. I'll keep on working. Now I know where I'm going.'

I do too, as a matter of fact. Tomorrow morning, I'm flying off to Nice, from there to Grasse and all the way up to the hills above the village of Cabris. I'm going to see where our brainchild will be born if it ever comes to term.

25

'This is where my father grew his lily-of-the-valley.'

Michel Roudnitska, a tall lanky man in his early sixties whose kind, slightly melancholy smile softens his ascetic features, has been walking me around the eleven-hectare garden landscaped by his father Edmond. Many pilgrims visit Sainte-Blanche, his family home, and Michel knows that this spot is the highlight of the tour. Delicate vegetal smells permeate the warm, damp air; a gossamer-thin haze refracts the light of the September sun. I am nearly hyperventilating in my quest to breathe in the same aromas the great Edmond Roudnitska must have done, as though the genius of perfumery could be assimilated through osmosis. In this intensely technical world, we are never very far from magical thinking.

I gaze reverently at the mass of sprightly oblong leaves jostling at the foot of a wall tangled with winter jasmine. The lily-of-the-valley patch isn't much of a sight at this time of year, yet it is the mother of all lily-of-the-valley patches, the one that got Edmond Roudnitska and Christian Dior on their hands and knees to bury their noses among its tiny white bells. This patch inspired one of the best fragrances of all time: Diorissimo.

Edmond Roudnitska grew lily-of-the-valley, the essence of which can't be extracted, to study it in all its nuances. His rendition of it wasn't a copy but an 'arabesque' that connected all the facets of the flower to the vernal landscape where it grew. In doing so, he later wrote, he expressed the yearning of a world just coming out of years of war and restrictions for something fresh, pure and green. A decade later, his Eau Sauvage would also become the expression of the zeitgeist, an entire generation's olfactory badge. Though it was marketed to men, the 'Wild Water' was promptly filched by young women who rejected the headier, more ladylike classic French perfumes. Michel Roudnitska recalls that as a student in Paris in the late 60s he would be engulfed in a cloud of Eau Sauvage whenever he stepped into amphitheatres. He also tells me, surprisingly, that Eau Sauvage was rejected by consumer panels: the head of Christian Dior Parfums launched it against the advice of his marketing team. No wonder Roudnitska was denouncing marketing tests as far back as the 60s.

'When I see the choice of a new composition subordinated to the opinion (oh so superficial) expressed by one hundred, two hundred or even five hundred women, when it must please millions of women in the world if it aims to be a great perfume, I wonder whether I'm dreaming. It is truly the triumph and glorification of irresponsibility, but it is also the negation of art and the stifling of talents,' he wrote in 1972, appalled that a whopping dozen perfumes had been launched that year and denouncing 'the veto opposed by the incompetents who have the power to decide'.

It is this fighting spirit I have come to commune with. Everything that Edmond Roudnitska stood for. Art. Creative independence. *Le beau parfum*. But though Michel and I have paused in front of the small oratory where his parents' and paternal grandparents' ashes rest, Sainte-Blanche is not just a place of pilgrimage, the cenotaph of a bygone era. The Roudnitska heritage lives on. The flame has been kept alive, it's being fanned, and it's catching. Art et Parfum, the company Edmond Roudnitska set up here in Spéracèdes after World War II, has opened its doors to perfumers who have broken free from the 'studio system'. The heirs of Edmond Roudnitska, if not in style, at least in spirit, have found a home here, Bertrand Duchaufour among them. Could the birth of Duende be placed under better auspices? My perfume may one day be part of the story of Sainte-Blanche and I want to experience its physical reality. I've also come to meet the man who's taking Art et Parfum into the 21st century.

'You'll see, it's a magical place,' Bertrand told me as I was arranging my visit. And it is. When Edmond Roudnitska found it in 1947 as he was cycling up the hills on the beaten-earth roads above the village of Spéracèdes, the land was nothing but rocky *garrigue*. 'I will make the rocks blossom and the birds sing,' he decided, and he had the phrase engraved on a stone arch above the entrance to Sainte-Blanche. Here, with his earnings from Femme, his first great success, he built a tall square three-storey house in a spare Art Deco style; the plant, another three-storey building that looks more like a residence than an industrial facility, stands a few hundred yards away. Here, he planted the garden where he took daily two-hour walks: its austere lines were a reflection of his aesthetics, says his son, 'with a balance between natural structures and human intervention'. After spending ten years in Tahiti, Michel settled in the family home with his wife and added his touch to the land, a Japanese garden where the human hand is imperceptible. Soon, a garden of odorant plants and an iris field will spring from a part of the estate that was never landscaped: it will hum with beehives to keep it alive, their pollination activity a symbol of what Sainte-Blanche stands for. This will be Olivier Maure's contribution.

When he joins Michel and me, a handsome forty-something with open features and the lilting accent of Provence, the garden starts buzzing, and it's not just the bees. Olivier Maure is an entrepreneur but he's also a believer, and almost as much a child of Sainte-Blanche as Michel: he was hired by the Roudnitskas when he was nineteen years old, came up through the ranks and rose to be the head of the plant, then its majority shareholder. He now helms Accords et Parfums, the production branch, and Art et Parfum, the creative branch. This is still very much a family operation, and as the vital, voluble Oliver and the tall thoughtful Michel lean to pick herbs or draw down branches so that I can smell leaves and blossoms, I feel as though I want to be part of it.

But it is of my long-distance friend Sandrine Videault I think as I climb the staircase of the Roudnitska villa, pausing to gaze at the paintings lining the walls. Sandrine came to Sainte-Blanche in 1992 as a business student to interview Edmond Roudnitska for her Masters dissertation on the luxury industry. When she left eight hours later, she was on her way to becoming a perfumer, and had found mentors in the Roudnitskas.

I gasp as Sandrine must have done when Michel leads me to the first-floor terrace, with its Ionic columns and wisteria hanging from the pergola. From this eagle's eyrie – the eagle that spreads its wings on the family crest, a symbol of Edmond Roudnitska's lofty views on his art – you can see the coast of the Riviera spreading its arms to embrace the Mediterranean: two hundred kilometres of coastline from the Bay of Nice to the Gulf of Saint-Tropez. In the muggy autumn air the view is hazy, but when the mistral sweeps the sky blue, Corsica rises from the sea. Here too it seems important to smell the things the Roudnitskas planted, and I do, leaning into the box hedge to sniff its green leaves (cat pee!) while Michel stoops to pick a sprig of rosemary and crush it between his fingers for me to inhale.

When we've settled into the blond-wood-panelled corner office where Edmond received his visitors and Thérèse worked, its windows revealing an even clearer view of the hills rolling towards Grasse, Cannes and Nice, I ask to experience the master's scents *in situ*. Despite having smelled them all their lives, Michel and Olivier dip strips for themselves and, together, we contemplate Roudnitska's 1949 Diorama, a lush fruity chypre with an almost animalic ripe-melon top note and leathery under-tones. The scent straddles the spicy, plum- and peach-tinged Femme and the sparer, more sparkly Parfum de Thérèse.

It is indirectly because of Le Parfum de Thérèse that I met Michel a year ago. In a review I'd done of Émotionnelle, a fragrance he composed for the American brand Parfums DelRae, I'd recognized a tribute to his mother's perfume, though the scent was sold as a reminiscence of DelRae Roth's first trip to Paris. I'd even guessed its working title: 'the Melon'. Michel was moved and wrote to explain that he had indeed composed it from his memories of his mother's scent, purposely refraining from consulting his father's formula.

Olivier fetches another of Michel's compositions, Noir Épices, which came out at the same time as Le Parfum de Thérèse in Frédéric Malle's original line-up.

'When you weigh it, it's a feast for the nose, because the sequence is perfect even in the vat. Each time you add a new ingredient, there's a balance', he enthuses.

'Oh? I didn't know that,' says Michel.

Weighing is the term used in the perfume industry for meas-uring and adding the quantities of materials specified in formu-las into the vat where the oil is blended. It can be an automated process but for the smaller volumes treated by Accords et Parfums it is carried out manually, especially since the formulas are very complex and many of the products are so costly they must be handled 'with care, feeling and delicacy', explains Olivier. To him, weighing a formula is like performing a score, and the Accords et Parfums team communes around the ingredients to make sure it will be performed without a false note.

'What's interesting when you make up large quantities is that you know straight off which perfumer you're working for. Just by weighing, you read the way they write.'

Perfumers not only have distinctive palettes but also their own way of harmonizing proportions. As each material is added, the structure unfolds one accord after another: 'You can smell the fragrance taking shape all the way from the garden,' says Olivier.

For instance, he explains, Bertrand uses a lot of rich, powerful, complex raw materials, not in an overdose necessarily, but not as a trace, like many perfumers do. He manages to harmonize these materials, to give them softness, roundness, 'almost as though he were drawing with charcoal. And you can tell he starts each of his formulas from scratch.'

I wonder what Duende will smell like when it is mixed: I hope I'll come back one day to follow its birth. I envy Olivier this daily intimacy with aromatic materials and with fragrances as they take shape, the depth of understanding it must yield, the sheer physical pleasure of being permeated by those beautiful smells ...

Olivier is just as talkative as I am and by the time our conversation winds down, the plant has closed for the day. I'm crestfallen: I'd expected to see the mixing of some fragrant cocktail right under my nose, perhaps of something I know. Olivier tells me there's nothing very spectacular to see. Still, he'll show me around, so we walk across to the plant. I stand in the gravel courtyard while he fetches the key and lifts a corrugated iron curtain to let me into a warehouse whose walls are lined with plastic and metal drums on shelves: litres upon litres of raw materials, both natural and synthetic.

Sourcing, buying and stocking them is the first service provided by Accords et Parfums, and a major one, as independents can't generate enough volume to buy large quantities. Accords et Parfums can count on a network of suppliers built up over sixty years to ensure sufficient high-quality stocks at a good price. Materials can also be purchased directly from local or foreign growers/distillers. Brokers play an important part in the process: they buy up large quantities of natural materials as they become available, even when composition houses don't have an immediate need for them. Say you miss the narcissus season or the patchouli crop in Indonesia is poor, you're done for if stocks haven't been built up in preceding years. Brokers have those stocks on hand.

There's half-a-million euros' worth of materials in this room but money's not what I'm thinking of. When you're used to sniffing a strip dipped in a dinky little phial, the idea of sticking your head into a vat of myrrh – a gem-clear, syrupy brownish-red goo wafting whiffs of mushroom – of holding enough jasmine concrete in your hands to smear yourself from head to toe, or of diving into the 250-litre stainless steel vat giving off smoky resinous fumes … It's enough to induce a terminal attack of kid-in-a-candy-shop indecisiveness. Olivier shakes me out of it by prising open the lids of several drums and letting me moan a little as I snort. He seems to get a kick out of smelling the stuff too: working with it for half a lifetime hasn't made him blasé.

The rest of the stock is in quarantine, which is to say that it hasn't been analysed and approved for use yet. We step into the next room, where the raw materials and perfume blends are tested for quality, chemical composition and compliance (naturals may contain allergenic molecules in varying quantities according to the season's crop, provenance or extraction process). These analyses ensure that suppliers haven't played fast and loose with their wares. For instance, they might add synthetic linalool, a molecule already present in lavender essential oil. This type of tampering, called 'commercial quality', ensures that a composition house will become dependent on the material provided by a specific supplier. The practice used to be widespread but has become less frequent over the past decade as composition houses equipped themselves with testing facilities.

The first analysis is carried out by a sophisticated piece of apparatus that takes years to fine-tune: the nose. Materials are also chemically analysed and submitted to gas chromatography, a technique that can read the breakdown of a material or fragrance. The gas chromatograph looks like an oversized microwave oven. You pump a bit of the material you want to analyse with a syringe and pop it into the machine, where it is volatilized in a hot injection chamber. As the different molecules go back and forth between the gas phase and dissolution in a high-boiling liquid, they do so at different speeds and in varying patterns. These patterns come out as a series of peaks and troughs on a graph. You can then read them with a piece of software. When you slide the arrow on a peak, a pop-up window tells you which molecule it is, with a percentage of certainty.

There's nothing more to see, so Olivier leads me back to the gravel courtyard and, while I admire the vista one last time, he tells me his plans for Accords et Parfums.

'We don't want to be a company that's done beautiful things: we want to use our history to go on doing beautiful things. And the idea is starting to create a buzz, because our way of seeing things is totally atypical. For artists, it's a way of expressing themselves fully: this is what's missing in the industry nowadays.'

But it's not just about art, he says: it is also about the pleasure of working together; about friendship. I don't know what Bertrand is doing just now, but he must be on the verge of deafness from the buzz in his ears. Olivier is convinced he's going to be huge.

We climb into the company van so that Olivier can drop me off at the dingy Hôtel des Parfums, surprisingly one of the few decent accommodations in the touristic town of Grasse, its rooms plastered with warnings to lock doors and windows at night for your safety. I realize I've forgotten to let him smell the two latest versions of Duende. It's the first time I show it to anyone connected with the perfume industry, but I figure I'm keeping it in the family. We're heading for a roundabout as I spray a bit of each on my wrists.

'I can't stand it! I just have to smell this now!'

Olivier lunges for one of my arms, pulls it towards him as he negotiates the roundabout with his free hand and ducks nose-first towards my wrist.

'I can totally recognize Bertrand's style here!'

Without letting go of my wrist, he asks me to hold out the other one and inhales robustly.

'Uh-uh. It's very good but … I shouldn't really be saying this, I guess … But … there's still work to be done, isn't there?'

We've reached the hotel. Olivier parks the van, gets out and holds out his own wrists, grinning.

'I've got to try them on too. May I?'

I spray his arms and he waves them about to dry them.

'Sorry,' he grins, 'the day's been muggy and I'm a bit whiffy.'

Before I know it, I've blurted out:

'Oh, please don't apologize: I *love* the smell of men …'

But what I really mean is: Olivier, I love the way your sweat smells, just like rye bread out of the oven.

On his skin, N°14 is breaking up pretty badly, the grapefruit and flint sticking out like shards of glass. I study both mods, all the while copping an olfactory feel – hey, why shouldn't I?

Still, it's time to let this nice fellow get back to his office: my visit has disrupted his routine and he's got a lot of work to do before returning to his family. We stand in the parking lot, still talking a mile a minute.

'Funny,' says Olivier, 'the first time I met Bertrand a couple of years ago, this is where I dropped him off, at this time of the day. He said "I think we'll be meeting again."'

I lean forward to kiss him on both cheeks – we've shifted to the familiar *tu* which means *la bise*, rather than *le* handshake, is now in order.

'Well then, I'll say it too: Olivier, I think we'll be meeting again.'

26

As I stalk the arcades of the rue de Rivoli on my way to the lab, Lou Reed is droning into my iPhone buds that he's waiting for the Man …

Today's the day I'm putting down my zebra-striped four-inch-heel stiletto. I've been waiting for the man for nearly six weeks. We met once after the August holidays and now it's late October. In fact, since we started working on the project six months ago, it seems I've done almost nothing but wait: Bertrand's been away from Paris nearly half the time on various promotional jaunts, sourcing trips and holidays, and when he *is* in Paris, there's always something more urgent on his agenda.

Granted, he's the hottest indie nose-for-hire at the moment. Clients eager to capitalize on his talent and reputation are beating a path to his lab; he's juggling several contracts and each time I see him, he tells me he's about to take on more. Fair enough: he's freelance, ambitious, eager to explore as many different registers as he can. But he's got no agent, no PR, not even an assistant to help him handle increasing demands on his time and creativity, and I'm wondering whether he knows how to pace himself. More than that: I'm worried. I like the guy, respect the artist, and want neither to burn out. The angel in me wants to protect him; the muse is whispering that his art must be nurtured, not exploited, not even by himself. But my demon *duende* is slipping dark, mouth-burning anger between my lips. The fact that I can't stand being pushed back into the crowd of those clamouring for his attention is a matter of pride, and since that can't be helped I've got to swallow it. Yet it's not just a matter of pride. What we've been doing *is* different from his usual gigs. The idea sprung up when our orbits intersected: spontaneously, gratuitously and, therefore, out of necessity. If there's a blood accord in Duende, it's because there *is* a bit of our lifeblood in the phials lining up on his shelf and their twins huddling next to my computer. But lately it's been running a little thin.

I'm not just waiting for the Man. I'm waiting for the *Moan*, the one I let out when a perfume hits me at gut level. The Moan hasn't come out, though that may be part of the protracted process of composition: dozens of mods before you can let go of any rational judgement and surrender to the beauty. Nevertheless, I can't help wondering whether we haven't taken the wrong fork somewhere. The last time we saw each other, Bertrand told me from now on I'd be seeing small tweaks, unless we wanted to change our course drastically.

And that's just it. We've been focusing on tweaks; narrowing the scope to technical details: safe ground for him, not least, I suspect, because it's a terrain where I have very little say and he can move quickly. Meanwhile, we're losing sight of the thing that brought us together. It's not just because he's been morphing into a star over the past few months, the acclaim and contracts giving him constantly renewed motives for distraction. I've been thinking about my conversations with artistic directors, especially Christian Astuguevieille and Pamela Roberts, who both worked with Bertrand, but also Serge Lutens and Frédéric Malle. They all said the same thing. Ask yourself what the perfume wants. Trust the perfumer. Give him as much poetic licence as he needs. But make sure he stays focused on the story.

It's time I heeded those lessons. I'm the one who should be keeping us on course yet practically all I've been doing so far is to mirror Bertrand's decisions. When he asked me whether I knew what I wanted, I couldn't answer, and that still smarts. When I mentioned that the few friends who'd smelled the various mods had all been saying, 'This isn't you,' Bertrand kept repeating that he was perfecting the floral note. We'd inject more sensuousness afterwards. What could I say? I'm not the perfumer; I don't know how it's done. But however it's done, I *do* know one thing: this is *my* story.

By the time I'm climbing the stairs to the lab, Nico is cooing, 'I'll be your mirror ...' into my earphones. Well, guess what, Mr D.? The mirror is about to talk back.

Of course, as soon as Bertrand greets me with a smile, whatever annoyance I'd been feeling towards him dissolves: he is nothing if not disarmingly likeable. It's only after you've sparred with him for a while you realize he's got a rock-hard core, some savagely protected part of him that can't be budged and won't open up. The hardness I see in the way he holds himself, legs slightly akimbo, ready to stand his ground; a density in his physical presence honed by years of tai chi. But Aries-versus-Capricorn head-butting would be unproductive: we'd both end up with a migraine. So as Bertrand downs a soup, a sandwich and a smoothie while I untypically pick at a salad I can't even finish – I've swallowed my rant and it's giving me a touch of heartburn – I steer the conversation towards my concerns about his career: flicking the cape, as it were, to attract his attention. He concedes I may have a point. I don't press it. I am not my perfumer's keeper.

Then we find our way back to our usual banter and I spill out six weeks' worth of ideas and anecdotes, drawing him back into that little world two people create when they've been working together for a while, and he laughs and teases me for being so talkative – 'You're in love with words, aren't you?' – but I know from the twinkle in his eyes he's enjoying the stories. For charm to operate, you've got to be charmed, and we're both charmers: that's our little unspoken war, I think. This may have been a game of dominance all along, one I've been letting him win out of sheer awe that this thing was happening at all. But underneath the comfort I feel chatting and laughing with Bertrand, that tiny nagging bite of anger is keeping me on edge. I've got a couple of things in my handbag that I'm waiting for the right moment to spring on him.

This time, I *do* know what I want. Or rather, I know what the perfume is asking for.

Up in the lab, Bertrand has two new mods to show me: numbers 15 and 16.

'I'm focusing on the incense-blood note, maybe stupidly,' he says as he's preparing the blotters. 'Maybe we need to work more on the base. Maybe this shouldn't be a luminous perfume like I imagined at the start, with an almost cologne-like freshness. Maybe we ought to work on a headier orange blossom note ...'

Good. He's already moving towards my ideas without my having to nudge him. It makes it easier for me to formulate my critiques. For one, the banana was too strong in the last mods. When I'm in a jasmine alley, I get something oilier, spicier. And I'm still stuck on the N°5 mod we rejected last May because it wasn't an orange blossom but a fierce, clove-laden lily. Re-reading my journal, I noticed that I was strongly drawn to it from the start and that I'd mentioned it several times afterwards. Yesterday, as I was testing mods 11 to 14, I sprayed on this fifth mod and found a quality to it that was missing from what Bertrand's been doing lately, something that's closer to the way I envision Duende.

When I mention this he just says, 'Yeah, OK,' before going back to the latest mods. He's still trying to work out the warm blood note he associates with '*duende*, ardour, desire, so many things ...'

'And if you want, we could work spicy notes back in, a little bit,' he concedes.

Indeed. This time I'm not letting him off the hook.

'Speaking of spices, I've got N°5 with me. Would you like to smell it again?'

'Gladly.'

I pull the phial out of my handbag and Bertrand dips in the blotters.

'Oh yeah!' he blurts out. 'It's really …'

'… really what?'

'Really woody. Really spicy. But you don't smell the orange blossom.'

'It's the warmth of it I loved.'

Bertrand falls silent for a couple of minutes. Then:

'It's a good thing you brought back N°5.'

It is? I'm feeling rather smug now, but also a little bit scared. What if I'm derailing the whole process? No. I'm pretty sure I'm right about this. Still, I hedge:

'I know it isn't the story, but …'

'But even so,' he interrupts me, 'the accord is beautiful. And it's powerful, isn't it? In fact, there's a whole damn story in it!'

We do a side-by-side between numbers 15, 16 and 5: the latter's green notes are different; they work better with the orange blossom, with slightly raw orange-tree-leaf effects.

'You were right to bring back N°5,' Bertrand repeats. 'There's a strong accord, even though it's not one hundred per cent in the story.'

Guess I'm no longer just a human blotter, then.

Bertrand gets up suddenly to rummage through his files and flips through his sheaf of formulas. The woody note in N°5 came from a material called cedramber, which he used in his church incense, Avignon. He'd dropped it in the following mods. Now I understand better why I was so drawn to N°5 – yesterday, I even sprayed it alongside Avignon, so it seems I got the connection. The green top notes in N°5 were composed of angelica, tagetes and cassis base. Tagetes, a type of marigold, has dried banana and daisy facets. Cassis (blackcurrant) adds raspy, fruity, sulphurous tones. As for angelica (the green bits in Christmas cakes are the candied stalks of the plant), I'm mad about it and so is Bertrand. It's got four different facets, he explains, a green dark resinous one that's close to galbanum with almost stem-like effects, a spicy one that's celery, a bit curry, almost hazelnut, and a musk note that is chemically identical to animal musk ...

'We'll put angelica back in,' he decides.

'And it kind of works well with the religious theme because of the name.'

He raises his head from his formulas.

'You know what? I'm going to go back to the base of N°5.'

'That means you'll be doing something completely different from N°16.'

'Well, yeah. I'm completely going back to N°5 with its spicy note.'

'You agree with me then?'

'Yes, because N°5 has more character, more body, more accords than N°16.'

Should I be pushing my luck? I've scored high marks for sticking N°5 back under Bertrand's nose – hey, Coco, could you be watching over me with your lucky number? As long as we're shaking things up, I might as well get out exhibit number two: an 80s bottle of Habanita. Yesterday, I tried spraying on a tiny bit next to N°12 and Avignon to see how they combined, but Habanita just gobbled up my whole forearm.

Bertrand pulls a face.

'The problem is you're bringing this up so late in the game.'

Sheesh. I've been talking about Habanita for months. It's called selective listening, darling: men are very gifted at that.

'I don't want Duende to smell like Habanita. It's gorgeous but it's not modern. What I *do* want is that burnished gold quality, that darkness you know how to work in so well ...'

'That Caravaggio chiaroscuro, as my friend Chandler Burr would say ...'

The *New York Times*' former scent critic, who considers Bertrand 'a living Old Master of scent', did indeed write that his Paestum Rose for Eau d'Italie was 'rich and filled with meaning like the intimate opalescent blacks Caravaggio painted.' Caravaggio is, along with Francis Bacon, one of Bertrand's favourite painters ...

'Exactly. That darkness has got to be in our fragrance. The boy in the story smokes brown tobacco. It's the smell on his fingers ...'

Bertrand sniffs at his blotter of Habanita, eyebrows knitted.

'I'm stunned.'

'By what?'

'By the natural tobacco effect. If this bottle is from the 80s they could probably still use tobacco absolute. Now to use a good tobacco that's not banned because of the restrictions on nicotine ... It's rough going.'

Is he sold? I suspect he's a little annoyed because, all of a sudden, Duende is going in many different directions, all of which interest him.

'That's a lot of information to take in all of a sudden ...' I half apologize.

'All of a sudden ... Can I associate all of it? I don't know. It could be interesting. But I'd have to rebuild the formula completely. Rework the elements one after another, try to associate ... How can I say? ... the *patterns* of the accords, the orange blossom note as it is in 15 and 16, the lily note as it is in 5 and the Habanita note as I smell it here ...'

He points to the bottle.

'The tobacco?'

'Oh yes, it goes with the story, *completely*! Tobacco, tobacco-stained fingers, and what have you!'

'Tobacco-stained fingers in my knickers,' I slip in, not wanting him to lose sight of the erotic aspect of the story.

'In your knickers,' he echoes in a get-your-mind-out-of-the-gutter tone, before adding thoughtfully: 'This can be a beautiful challenge.'

'So we're going for that?'

'We're going for that. I'm sure we can get there. Now I have to break everything apart … We're starting from scratch!'

'Am I making you suffer with this?' I say, a bit hopefully.

'No.'

'Good, because that's not the point.' ('Yes it is,' my demon *duende* whispers.)

'I'm not afraid.'

'*Même pas peur!*' we utter simultaneously, laughing. It's one of Bertrand's pet expressions, something kids say in the schoolyard: 'Not even afraid.'

'In fact we're not really starting from scratch,' he says. 'We'll dismantle the puzzles, scatter all the pieces, take some from one puzzle, some from another, and re-assemble the perfume in a different form … bits of story that will shape another story.'

'What a plot twist!'

'It's a plot twist, but that's often how things go when you're making a perfume.'

We take a breather by going over my write-up of the session during which we discussed incense: this has been our process ever since I started the journal. I add bits of research along with my thoughts to the transcripts, which allows Bertrand to follow the course of the development and put it into perspective. In this particular case, he's both pleased and astonished to find out that his intuition about the equivalence between blood and incense is rooted in several myths, as well as in Catholic liturgy. He re-reads our conversation about his own obsession with the note and stares at the page for a while, nodding.

'You're right, you know … My creative process *is* a kind of therapy for all the suffering and frustrations of my childhood … I wasn't mistreated, that would be an exaggeration, but our super-strict Catholic upbringing was quite oppressive. We've tried to break free from it and mostly we've managed to, at least the youngest siblings. Perfumery was one of the best ways to act it out.'

I am suddenly reminded of Serge Lutens' droll, indignant tirade against bespoke perfumes – 'I'm not a psychoanalyst!' During lunch, I quoted it to Bertrand, who wholeheartedly agreed: his own experience as a bespoke perfumer ended abruptly after spending a whole day with a client. 'At the end, I knew his life story but nothing about his tastes in perfume!'

Bertrand and I are far from psychoanalysing each other, though smells have a way of triggering sudden disclosures. But somehow it seems right that reading through the incense session and going back to the painful memories it summons – reopening the wounds of the incense tree, as it were – should come just as we're dismantling what we've done up to now. Not just the formula, but the way we work together. Not quite a power shift, but greater exposure. Any creative process, when it is true and sincere, is like baring yourself to the bull's horns. This is the first time that I lower my arms to bring the cape, and thus the horns, closer to my body … Though I did it in a roundabout way, showing him the things I wanted so that they would seem obviously right to him rather than asserting my will, I've exposed my desire, what I wanted for the perfume, and thus taken the risk of his saying no: I still haven't shaken the feeling he could buck at any moment. A Capricorn goat may not be able to gore you like a bull, but he can knock you out cold or throw you out of the ring.

Still reading through my write-up, Bertrand gets to the point where he suggested adding an ash note.

'Why don't we put it in? The ashes of Christ, cigarette ashes … It's all part of Holy Week, after all.'

Well, no, actually. Perhaps he's willed his Catholic education out of his memory; still, he must have known at some point that Ash Wednesday is the first day of Lent, a full forty days before

Easter and therefore, not part of Holy Week at all. The ashes are obtained by burning the palms of the preceding year's Palm Sunday. They are a multi-secular symbol of penance used as far back as Ancient Egypt, as well as a reminder that we are dust and will return to dust. What they are definitely *not* are the ashes of Jesus: that he resurrects is kind of the whole point of Easter, not to mention Christianity.

But I'm not about to dive into a theological debate with Bertrand because, right now, he's doing something I've never seen him do, drafting some kind of flowchart. As I peer into it, I realize he's laying flat the structure of Duende.

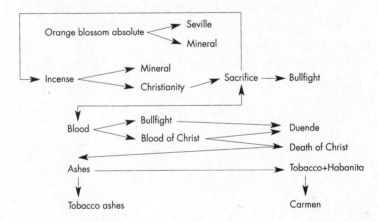

'Do you often do this?' I ask him.

'Yes,' he mumbles, still scribbling and drawing arrows.

I let him finish the chart before speaking again:

'It's as though something had clicked all of a sudden. As though you'd rethought the whole thing out.'

'What's crazy is that I've also rethought the whole formula.'

'Well, that's exactly what I wanted to do with you today. Shake things up.'

'This isn't about destroying what we've done. Absolutely not … it's just …'

'… clarifying our ideas …'

'You know, creating a perfume is a labyrinthine process. At one point you say, "Shit, I've taken the wrong turn," and you have to turn back a bit to take the right direction.'

'At least the fact that there are two of us helps us get some perspective.'

'That's essential.'

Six months and eight sessions into the development of Duende, it seems I've finally taken on my role as a creative partner. I've been wandering in the labyrinth, but now I've found the red thread. No longer a muse but Ariadne leading Theseus towards the Minotaur …

The bull. The blood. The wounds of the incense tree. 'The aid of the *duende* is required to drive home the nail of artistic truth,' writes Lorca.

'I'll write down the formula right away before I forget anything,' says Bertrand.

'You won't.'

'I won't.'

27

Three bottles of Habanita are now sitting on the shelf above Bertrand's desk. The first contains the umber dregs of an extrait I bought in the 80s. The second is a black glass bottle of eau de toilette with a Lalique frieze of nymphs, the one with the tobacco note Bertrand loved so much. The third was sent by an American friend who stumbled on a stash of vintage perfumes in a second-hand shop near Austin, Texas. Miraculously, this particular bottle of Habanita came with its date of purchase: a page of the 2 July 1948 issue of *Le Monde* had been wrapped around it so that it wouldn't be jostled in its box. Now yellowed and brittle, it announces the adjournment of the Frankfurt conference convened to give West Germany its constitution and the expulsion of Yugoslavia from the Communist bloc: news of a bygone world in the throes of post-Yalta reconstruction, on the day the tiny bottle started on its journey from Grasse to Austin. And now, over half a century later, the crystal time capsule has made its way back to France. I brought it to Bertrand two weeks ago, a couple of days after our momentous 'coup de théâtre' session.

He has drawn his inspiration from all three bottles to compose different 'Habanita bases' on an accord of tobacco, oak moss and cistus, a resin with amber and aromatic facets. He's also put in vanilla and coumarin, which has almond, hay and tobacco effects: they are two of the main notes of Habanita.

I lean towards the blotter dipped in base 2 and find myself on the verge of the Moan for the first time since I smelled Duende 5. It is a balsamic caress: sweet, powdery wafts wrapping the richest tobacco. In this concentrated form, one you'd practically want to lick off the blotter, it smells like Bertrand's Vanille on steroids, unsurprisingly, since the two formulas have several materials in common.

But as soon as I have a whiff of base 3, I *do* let out a little gasp. I *know* this. Of course, I know *rationally* that Bertrand has based the formula on Habanita, but this grabs me at another level: I've *lived* here, *inside* those notes. The sense of familiarity is even more vivid than when I smelled my old bottles after pulling them out of the closet where they'd been banished. It's as though by paring the structure down to its load-bearing walls, Bertrand had managed to display Habanita's quintessential beauty, to make it more *legible*.

Base 3 is a perfume in and of itself, though Bertrand says he's only added three materials to the formula of base 2: musk ketone, methyl-ionone (a molecule that smells of iris, violet, wood and tobacco, with a powdery effect) and aldehyde C11, which smells of snuffed candle. Because these materials became ubiquitous after the success of Chanel N°5, base 3 conjures half of my vintage perfume collection: it is the backbone of classic-era perfumery.

As Bertrand had said he would the last time we saw each other, he's concentrated on building up the tobacco and orange blossom accords separately. In the latter, he's added beeswax, I'm happy to note. We'd discussed it last spring and then it got lost in the fray, though it *is* a major element of the story: the candles carried by the penitents and illuminating the floats bearing the statues of the Virgin and Christ. The beeswax is also logical in purely olfactory terms: it acts as a bridge between the orange blossom and the tobacco because it has tobacco facets, while tobacco, though it's more acidic, has slightly waxy effects.

On the other hand, he's scrapped the banana by eliminating the ylang-ylang and cutting down on the jasmine absolute. That's good, since it didn't belong in the story, but he had his reasons for including it in the first place: the fruitiness was meant to cover up the blood and raw meat facets of the incense, so that he could work a higher dose of it back in. I can detect the incense, but not as much as I'd like. Bertrand says that, when he adds the wood base, plus the resins and tobacco of Habanita, that little bit of incense will be boosted.

The new orange blossom comes off as quite metallic since Bertrand has kept the rose oxide from mods 15 and 16 to conjure the blood accord, but there's also something that makes me wonder whether he's skipped a shower. I'm about to drop my pen to the floor so I can lean towards him and have a surreptitious sniff when he informs me he's upped the costus.

'Oh, and here I was thinking it was you!'

Now I'm getting a pretty animalic note from the Habanita base 3 as well ... Bertrand's mouth curls into a sly little smile.

'I put in African Stone.'

I let out a delighted hoot. When I'd given the African Stone to Bertrand, I'd asked him whether it would suit Duende and he'd agreed to try it out. Next thing I knew, he'd used it in another project. I was happy to have fed his inspiration, but slightly cross that someone else would reap the benefits of my little gift. As it turns out, Olivier Maure from Art et Parfum, with whom I'd discussed the possibility of using such a material when I'd gone to visit him, sent Bertrand a sample from a different source. He retrieves it from his refrigerator, unscrews the lid and hands it to me gingerly.

'It's phenomenally powerful ... don't spill any on yourself!'

'Or I'll be followed by all sorts of animals, including the human kind ...'

The late-afternoon November gloom is falling over Paris: time to move on to mods 17, 18 and 19, which combine different proportions of the new orange blossom base and the Habanita base 3. I'm squirming with anticipation on my swivel stool.

My smile fades when I smell mod 17. The aldehydes in the two bases have ganged up on the other notes and taken over in a huge blast of hot iron steam. Bertrand shakes his head.

'OK, forget 17. It's too old-fashioned, too Habanita.'

I guess he had to rebuild it to understand it from the inside.

'Of course,' Bertrand says, 'but it's really interesting to integrate this base into our story. It's part of it. That's why I went back to it last time.'

Aren't we rewriting history a bit here? *I* brought it back. But never mind.

'And what about N°5?' I ask him. 'I'm really hooked on it.'

His gaze softens.

'You like it. Good … that's good,' he says almost tenderly. 'We'll get back to it.'

When you smell formulas that have a lot of common ingredients, it's what's different that sticks out, as though your brain blocks out the identical bits. In this case, in contrast with mods 18 and 19, mod 16 comes off as banana jam, which it wouldn't on its own. But the new stuff isn't working out either: the Habanita base, like a hungry enzyme, gobbles up the orange blossom even in mod 19, which has the smallest proportion of it, one fifth. Clearly, classic-era forms like the one Bertrand rebuilt are incredibly powerful, especially Habanita – it *was* marketed as the most tenacious perfume in the world. This is because there are only base notes in the formula, he explains.

'It's the base notes that give the direction, no matter what they say. You'd almost have to work on the base notes first to know exactly what a perfume's going to become.'

That must be why people who are used to wearing classic perfumes don't find contemporary ones strong enough. As soon as you stick in the stuff that's in the Habanita base, it crushes everything. But the experiment is interesting, because it helps me understand what perfumery used to be like.

'First and foremost, you worked on the base notes,' Bertrand repeats. 'What you put into the heart and top notes was almost solvent. That's actually what Jacques Guerlain used to say: that he used bergamot as a solvent, in sufficient quantity to finish his formula. That's why there's as much as thirty per cent bergamot in Shalimar. To him, it didn't mean anything. His perfume was done. Still, his bloody bergamot *did* something. It had an *action*. But it wasn't part of the formula.'

I love it when Bertrand talks about the lore of the industry. Like couture, perfumery is a trade where oral tradition is part of the learning process; when he was an apprentice, anecdotes about historical figures were swapped in the labs by people who'd actually known them. Bertrand's mentor was Jean-Louis Sieuzac, who started out at Roure before going to work for Florasynth, the company that took Bertrand on as a trainee when he first came to Grasse. Roure had its own school of perfumery, where the pedagogical methods devised by the head perfumer Jean Charles were used to train several generations of perfumers, including Bertrand, indirectly, via Sieuzac. And, as a beginner, he had to reproduce 'by nose' a great many classics, something student perfumers still do, so that of course he knows classic structures inside out.

It just so happens that the same friend who found my 1948 Habanita sent me the English translations of some articles by Jean Carles, written for an industry magazine between 1961 and 1963 and published in William I. Kaufman's 1974 *Perfume*. When I mention this to Bertrand, he tells me that this very book was one of the things that made him want to become a perfumer, though when he first read it in 1982 it had become so rare he was never able to buy his own copy. So it makes perfect sense that

Bertrand is echoing Jean Carles: 'As indicated by their name,' the latter wrote, 'the *base notes* will serve to determine the chief characteristic of the perfume, the scent of which will last for hours on end and will essentially be responsible for the success of the perfume, if any.' To Carles, the heart notes were just 'modifiers' meant to cover up the unpleasant facets base note materials gave off initially. Top notes were designed to draw in the customer, 'with or without reason, as in no case can the top note be the characteristic note of the perfume.'

Today, top notes are what perfumers focus on, at least in the mainstream, in a complete reversal of the Jean Carles method. As consumers seldom take the time to test a fragrance throughout its entire development or even to try it on skin before buying it, perfumers compose things that give off a seductive blast in the top notes and that smell good on cardboard. Little effort or budget is wasted on the base notes, which is why they often end up smelling so nondescript and generic, a mishmash of woody notes smothered in laundry musk.

For the time being though, our problem is just the opposite one: base notes so strong they take on a life of their own.

'OK, let me try out just one last thing before we wrap up,' says Bertrand.

This time he adds only one tenth of the Habanita base to the orange blossom. As soon as he's shaken the little phial, he dips in the blotters. The blend is likely to be a bit of a mess: to really judge a perfume you'd need to let it macerate at least one week after you've added alcohol to the oil, though one month would be better. But it'll have to do.

This time, the orange blossom, incense and blood accord stands its ground.

'Still, even if we play with the two bases we'll have to add other products. We're only up to thirty, thirty-five materials, so the formula won't be getting out of hand. There's something to go on here!' Bertrand concludes.

* * *

Of course I can't wait one week to try out Duende 20 on skin. The next morning I spray it on my wrists, hair and chest. Sticking my nose inside my sweater is a good way of getting a concentrated blast, so it's more effective to assess the form of the fragrance than gluing the wrist to the nose: a bit like stepping back from a large painting to see the composition rather than the fine details.

But however I smell it, Duende 20 is going all over the map. When it's just moving out of the top notes, the aldehydes tend to stick out, which makes it smell like a candle. When the orange blossom struggles out I find it a bit too soapy. And then the mineral effect literally explodes, giving off a slightly burnt-hair stench. But when I sniff my wrist, I don't get that note at all: it's all sweet tobacco, musk and vanilla. It seems a monster has shambled out of Dr Duchaufour's lab … And then an email pops into my inbox.

> Denyse, I'm afraid I'll have to cancel our appointment tomorrow. The mods I made after you left are crap … I'm even starting to wonder whether the orange blossom and incense accord is really relevant. I'm down in the dumps.
> B.

I call Bertrand back immediately. He's not picking up so I leave a message.

The train is rumbling into Daumesnil station when the Farfisa organ of 'Ninety-six Tears' tinkles from my bag half an hour later. I can barely make out Bertrand's voice in the station, so I run back up the stairs into the icy dusk of the place Daumesnil to huddle on a bench. I'm sick at the idea of having tampered with Bertrand's creative process, yet oddly calm. We can't both be freaking out. I shift into crisis-control mode.

The new ideas aren't working out, he says, and he's tired of running round in circles. The Habanita accord smells too old-fashioned. He tested Duende 20 on his skin the day after our last appointment and it was awful. Then he tried adding the woody, spicy lily and it didn't work out either. He hates the costus. Maybe there shouldn't even be actual incense in it. Maybe he should substitute other materials.

OK, if the Habanita accord is too invasive, I say, we'll ditch it. It had to be tried out. Don't worry. There's time. Let's get together soon and just sit down and smell everything again quietly, you don't have to come up with anything new, we'll just take stock, think things over, now that I'm more of a partner, take advantage of that, rely on me.

By the end of the conversation Bertrand is considering solutions, such as streamlining the formula of Duende 20 to remove the old-fashioned elements; I suggest trying out the orange blossom accord with Duende 5 but without the Habanita effect. What I want is the sensuousness of it. It needn't be literal. I'm still convinced we needed to shake things up.

Of course that's also the problem, I think as I hang up and make my way back to the station, still rattled. I've become more of a partner but I'm also more intrusive than any client of his ever is. I *am* derailing the process. Possibly asking for what *can't* be done. Things that won't work together within his style or won't work together, full stop. I've never realized until now how tough it must be to pull off this project. Bertrand is working under a triple constraint: interpreting my story as faithfully as possible; not repeating himself within a body of work of nearly fifty perfumes, several of which feature an incense note, and one of which is an orange blossom soliflore; coming up with an accord that's different from everything else on the market, all the while juggling rich, complex materials that sometimes interact in unexpected ways, despite his in-depth knowledge of them.

Still, I can't help thinking some good might come out of this creative snag, frustrating as it is at the moment, if only because I've found out how much he cares about the project. That he trusts me in moments of doubt. And that I trust him now more than I ever have, precisely *because* of this moment of doubt.

Bertrand, when you had the intuition about the orange blossom and incense accord, it was immediate, obvious. I have absolute trust in that intuition, that of a great perfumer. And in the accord, *because it exists in reality.* You'll get there, and your perfume will be heartbreakingly beautiful. Actually, you know what? *Même pas peur.*

D.

28

Is there an uglier church in Paris than La Madeleine? Though the demented meringue of the Sacré-Coeur in Montmartre looks like a pastry chef's idea of Byzantium by way of Disneyworld, at least it's got a certain hysterical élan. But this pompous faux-Roman temple looming sullenly at the end of the rue Royale doesn't seem to know whether it is the National Assembly, the National Library, the Stock Exchange or the Opera. In fact, it almost became all of those things between 1763, when Louis XV laid the cornerstone, and its inauguration by Louis-Philippe in 1845. There was a project to turn it into a temple to the French Revolution, then to the glory of Napoleon's Great Army; in 1837, there was even talk of making it Paris's first railway station. In the end, it was returned to its original destination as a church but, frankly, there couldn't have been a worse mausoleum to house the relics of Mary Magdalene. It is as though the triple chastity belt of snarling traffic, cast-iron railings and Corinthian columns had been expressly designed to contain the overflowing femininity of the sexiest saint in history: stiff male virtue, whether it is revolutionary, imperial or bourgeois, rising up to keep the erstwhile courtesan in her place.

* * *

I've stepped into La Madeleine on a sudden impulse after realizing I'd never once set foot inside despite living in Paris for over two decades, an oversight that can only be explained by its ugliness because, for a miscreant, I've certainly spent a great deal of time in churches. When I was travelling around Europe with my parents, we didn't pass one without at least peering into it. My mother and I would sprinkle ourselves with holy water of dubious bacteriological status and light candles, dropping pesetas, liras, guilders or francs into tinkling brass boxes, setting Europe ablaze with our wishes.

I've never entirely shaken the candle habit and, as Bertrand is tackling his *duende*, it seems like the right time to have a word with Mary Magdalene. After all, she *is* the patron saint of perfumers, as well as glove-makers and apothecaries (a logical extension of her speciality since both of the latter tradesmen were the original perfumers), but also hairdressers, gardeners, penitent sinners and converted prostitutes.

Before he decided to build an expiatory chapel in 1815, King Louis XVIII considered consecrating La Madeleine to the memory of his brother Louis XVI and sister-in-law Marie-Antoinette. The match between the tragic queen and the patron of perfumers, hairdressers and harlots would have been particularly fitting: in the slanderous pamphlets that circulated before the French Revolution, Marie-Antoinette was berated for her love of luxury and her unbridled lust, which sucked the country dry of its riches and vital forces.

Both lust and luxury, as it happens, are coupled in the same Latin word: *luxuria* is one of the seven deadly sins and one to which women, given their weak, vain nature, were thought more vulnerable. But if the fashion-mad Marie-Antoinette could have stood accused of spending too much on her baubles, she was never the sexual ogre libels made her out to be. Of course, that wasn't the point. What spurred on the slanders or, more specifically, their pornographic nature was an age-old fear of female sexuality. The lure of beauty, set off by costly and deceitful adornments, could lead men to material and moral ruin but, more frighteningly, suck them into a vortex of erotic voracity. A man's desire waxes and wanes. But how can a woman, whose pleasure is never certain and whose receptive capacity is potentially infinite, ever be controlled? And what then could be more terrifying than a queen in the throes of *luxuria*, whose power is unchecked by an impotent husband, as Louis XVI was thought to be? To the authors of the pamphlets, Marie-Antoinette was but the latest in a line of lustful queens that started with the wife of the stuttering idiot Emperor Claudius, Messalina, said to prostitute herself in brothels out of wantonness. The violent and fantastic nature of the accusations was directly proportionate to the terror aroused by the erotic power of women.

The fate dealt out to Mary Magdalene by the Catholic Church may bear witness to this fear. She started out in the Gospels as a beloved disciple of Christ. She ended up a whore, albeit a penitent one.

Perfume was part of her story from the outset. Her emblem in Christian iconography is the *alabastron*, the vase of alabaster containing precious aromatic substances. And her earliest known representation, in a Syrian fresco dated AD 232, is as one of the *myrrhophores*, the women bearing myrrh to Jesus's sepulchre to embalm his body, as the art historian Susan Haskins explains in *Mary Magdalen: Myth and Metaphor*. In Mark's Gospel, she is the first, along with Mary mother of James and Salome the Virgin's sister, to discover the sepulchre open and empty. She is also the first to see Christ resurrected in the garden of Gethsemane. She bears the news to the other disciples, thus becoming the first apostle (the word comes from the Greek meaning 'messenger, person sent forth'), in fact the apostle of apostles. She is also identified by Mark and Luke as the woman 'out of whom went seven devils', and those devils may well have sealed her fate: surely they were embodiments of the seven deadly sins, and surely *luxuria* must have prevailed in a woman.

In the four canonical Gospels, Mary Magdalene barely utters a word. But she plays a very different role in the apocryphal Gospels of Philip, Thomas and Mary (the latter supposed to be the Magdalene) suppressed by the Church as it eliminated its competitors by labelling them heretics. In this parallel tradition, Mary Magdalene is the bearer of teachings Christ imparted solely to her.

As the Church gradually pushed women back from the positions of ministry they had held in early Christianity, Mary Magdalene relinquished her initial status as the apostle of apostles: to all ends and purposes, she was gagged. Her morphing into the figure of the penitent whore sprung into motion in the third century, as she was gradually conflated with two other female figures in the Gospels, whose common point with Mary Magdalene was perfume.

The first is an unnamed 'sinner in the city', supposed to be a prostitute or an adulteress. She washes Jesus' feet with her tears, wipes them with her hair and anoints his feet with fragrant ointment in the house of Simon the Pharisee, in a sensuous gesture that almost seems to be the feminine equivalent of baptism. The second is the sister of Martha and Lazarus, Mary of Bethany, who also anoints Jesus by pouring precious spikenard over his head in a pre-figuration of his embalmment rites ('against the day of my burying hath she kept this').

It was around this triple Mary Magdalene that the legend formed throughout the Middle Ages. She became the heiress of the castle of Magdala, descended from kings, and the bride of John the Apostle, who abandoned her immediately after their wedding at Cana to follow Jesus. Out of spite, she threw herself into prostitution, adorning herself with jewels, rich fabrics and perfumes the better to flatter her senses and entice her lovers, until she saw the light, repented her past sins and followed Jesus too. One version of the legend has her living out the rest of her days in a grotto in contemplation and penance, having forsaken all worldly goods, clad only in her long tresses: this time, she was conflated with the 5th-century Mary of Alexandria, a former courtesan who had retreated to the desert of the Holy Land.

As a penitent harlot, Mary Magdalene was invaluable to the Catholic Church. She symbolized the redemption of Eve and was an easier female figure to identify with than the Virgin Mary. If Jesus could forgive a whore, take her into his fold and even grant her the privilege of being the first to see him after his resurrection, then any sinner could be saved, even women who, like her, had given in to *luxuria*. And thus she became the protectress of reformed prostitutes in the charitable institutions founded to save them from their life of sin. Not to mention a favourite subject for painters. Her penitence in the grotto, clad only in the unfurling waves of her golden hair, pert breast darting out between vine-like tendrils, was an excellent excuse to paint beautiful naked women in the guise of religious art, especially after Renaissance artists revived the tradition of the classical nude. But not everyone was fooled. Susan Haskins quotes a Florentine nobleman, Baccio Valori, who told Titian that his Magdalene in the desert, though she had been fasting, 'was too attractive, so fresh and dewy, for such penitence'. The canny Titian 'answered laughing that he had painted her on the first day … before she began fasting'. By the 18th century, most of these penitent Magdalenes, some of which were portraits of famous courtesans and royal mistresses, had turned into thinly disguised erotica.

When Sister Aline tackled her story in my high-school catechism class, the Magdalene was also an excellent excuse for teenage girls to talk about sex. Sister Aline was pretty cool about that. This was the 70s and Quebec had thrown off the shackles of the Catholic Church to fast-track itself into nationalism, feminism and the sexual revolution. But I don't remember Sister Aline telling us that the Church had cleansed Mary Magdalene's reputation in 1969. The unnamed sinner and Mary of Bethany had resumed their discrete identities, which the Orthodox Church had always staunchly maintained: Mary Magdalene was no longer the 'penitent saint' but one of Christ's disciples. Perhaps Sister Aline thought of this as a demotion and preferred the sexy Magdalene of her childhood?

I know *I* do. The woman whose 'sins, which are many, are forgiven; for she loved much', is the one I sought out in the churches and museums of Europe, from Bernini's *Penitent Mary Magdalene* in the Chigi Chapel of the Siena Duomo, marble made swirling ecstatic flesh, to *Christ and the Magdalene* in the Musée Rodin, of which the poet Rainer Maria Rilke wrote 'like a flame tormented by the wind, [she] tries to embed and hide the ineffable suffering of this so greatly loved body in her own broken love.'

It is to her that I have come today at La Madeleine. To the fallen woman who bore perfumes, and who has ever since fascinated men and women into spinning her myth until it was as inextricably woven with passion and sensuality as her hair was around the feet of the Saviour. To the woman whose sorrow was as luxuriant as her sins; whose tears were sweeter than the essence roses yield in distillation, 'sweating in a too warm bed' in the 17th-century poem by Richard Crashaw. Who could know more about *duende* than the Magdalene, who found her beloved arisen from the dead only to have him refuse her embrace? In Latin, his words to her in the Garden of Gethsemane, *Noli me tangere*, mean 'Touch me not.' But the original Greek in John's Gospel, *mê mou haptou*, could be better translated as 'Stop clinging to me' or 'Cease holding on to me.'

As I watch the wick catch and the flame flicker feebly in the chilly dankness of the cavernous church, I wonder whether 'Stop clinging' couldn't be my own motto. Perhaps pride supersedes *luxuria* in my soul? I neither cling nor let myself be clung to. My body *has* been touched, and my heart clawed bloody by many hands as I tore myself from embraces or threw myself into arms that drew me in before pushing me away, as I knew they would. So I have forged myself an armour – remove the 'r' and you get *amour* – out of words and adornment: the leather and silk, the black that frames my eye, the seashell curl of my lip, the velvety powder that stretches over my face like a veil. The stilettos punching patterns in the ground while their red soles beckon, in a 21st-century equivalent of the soles of courtesans' sandals in Ancient Rome, which printed advertisements for their wares on the sand they trod.

'*Sans talons hauts, on a le cul bas*': 'Without high heels, your ass is low' was the bit of advice passed on by Mistinguett, who reigned in French music-halls from the 1900s to World War II, to her junior, the raspy-voiced Arletty. I've heeded Mistinguett's advice and freed my *cul* from the laws of gravity. Having done so, I am forever at risk of falling – a grating, a pothole, a puddle will trip me up. In truth, I tell you, I am a fallen woman just waiting to happen ...

Whatever I believe or disbelieve, it all comes down to this: fallen woman. I may have cut loose from my Catholic upbringing, but I am still a Catholic by culture, the Virgin and Penitent Whore seared into my psyche. Is it any surprise that my path was bound to cross the Magdalene's often? That I've ended up dedicating myself to the art she protects? And that I uphold the heretic tradition of the Magdalene by revealing the Gnostic secrets of perfumery?

I come bearing perfume, but I also bear words. Touch me not, then, but follow my steps, hear my voice, see my smile, bear my gaze, surrender to my perfume. It is my perfume that calls you and draws an invisible bond between what you breathe and what I exhale.

The Magdalene first beckoned to me in the late 80s, when I was invited to watch the *rejoneadora* Marie-Sara fight bulls on horseback in the village of Saintes-Maries-de-la-Mer, in the Camargue. This was where, according to tradition, Mary Magdalene landed after being cast off in a rudderless boat by the pagans. It is tempting to think that from there she went off to kick-start the perfume industry. But she didn't make a beeline for Grasse:

according to *The Golden Legend*, she converted the pagans in Marseilles before retreating to a grotto in the Massif de la Sainte-Baume to spend the rest of her life in contemplation and mortification. Again, it is tempting to imagine that the mountain took its name from her association with aromatic materials. But though *baume* is the French word for 'balm' or 'balsam', in this case it is derived from the low-Latin *balma*, which means 'grotto'.

At Saintes-Maries-de-la-Mer, Mary Magdalene's cult has been superseded by that of Sara-la-Kali, 'Sara the Black', who is said to have been either the servant of Mary Jacoby or a local noblewoman who helped the Christian refugees, and is the patron saint of the Gypsies, who make a pilgrimage to Saintes-Maries-de-la-Mer every year on her feast day. But though it was Sara the Black's image that was plastered all over the village, it was of the Magdalene I thought as I was carrying on a tipsy late-night conversation with Luis Rego in a bar near the small village bull-ring. Luis, a Portuguese-born singer and actor in slapstick comedies, was a soulful, serious man off-stage. I'd been up to no good in Seville and the stories had somehow reached Luis – or was I the one who told him? With the arrogance of a young woman – and a slightly miffed one, as Luis showed no sign of being aware of my charms – I boasted that, when I sinned, my sins were great.

'Don't be so pretentious,' Luis answered. 'Your sins are insignificant on the scale of things.'

I stood corrected, but not cured of my Magdalene complex. More often than not, as Mary Magdalene was to the Virgin Mary, I've been the Other Woman. At least, that's what I was the last time I paid a visit to the sinner-saint.

Monsieur and I had just spent the night at L'Espérance, a luxury hotel just outside Vézelay, after gorging ourselves on the chef Marc Meneau's truffled *poularde de Bresse*. When we left in the morning, I asked him to go through the town so I could pop into Sainte-Marie-Madeleine, the Romanesque basilica set atop a hill, where it was claimed the saint's relics had been found (you could probably make up at least five Mary Magdalenes with the remains boasted by various churches). Monsieur's mobile phone rang as he parked in front of the basilica, so I went off on my own to take a few steps up the nave. A small choir of Vietnamese nuns was rehearsing in one of the chapels. I'd already been there with the Tomcat as we were driving back from our honeymoon in Italy, but at that moment I wasn't thinking of either of my men, the one I was still married to or the one who was married to another woman. Just surrendering to the light-headedness that comes from an evening drowned in wine and a few hours' sleep at dawn; sated flesh dancing in a beam of sun under the tall graceful Romanesque vaults. That feeling of lightness, the soaring of a body *redeemed* by beauty – not my body's beauty, but the beauty surrounding it, lifting it, carrying it upwards – was all that mattered. That day too I'd lit a candle, but as I hadn't repented I asked for no favour: I wasn't expecting the saint to bend the rules.

And now, here in La Madeleine, once I've dropped a coin in the alms box, I ask for none either. My business with Mary Magdalene is strictly professional. I ask for the perfume born from my memories of the celebrations of the Passion of Christ – of his night in the Garden of Gethsemane and of Good Friday when Mary Magdalene wept at the foot of the cross – to be as beautiful as I've ever dreamed it.

That's a prayer I'm sure she can answer.

29

The Roadrunner had a face for radio but a voice whose wavelengths could travel from your ears to the soles of your feet, pushing all the right buttons along the way. We entertained the kind of flirt the French are so adept at weaving into their professional relationships, at least when they know how to handle their native tongue in all its titillating subtleties. He'd been wounded by a sniper shot in Bosnia and was taking a break between war zone assignments by doing a radio series on historic Parisian hotspots of an entirely different kind. I was his guest expert and had been digging up the addresses of some of the brothels that had been closed down after World War II. We'd buzz the intercoms and interview the current occupants to find out if they knew anything about their building's former function.

The Roadrunner seemed pretty knowledgeable about women, so when he sprang the question on me in the same tell-me-everything-my-child tone he used on his interviewees, I skipped a beat before responding, unable to decide whether he was kidding:

'Do women fake it?'

Well … of course.

Then he wanted to know whether *I* did.

I did. I have. I do. Unless a man puts his hand to my heart, I'm pretty sure none has ever been able to tell that the back-arching, limb-twisting spasms and moans were just part of a show ... once in a while. I'm not above stroking a man's ego as deftly as anything else that comes to hand. But this time I'm not faking it. There's no need to. There never is, with perfume.

The sigh comes unprompted as I'm sniffing my wrist, sitting on a swivel stool in Bertrand's lab. Aesthetic judgement plays no part in this type of frisson; it's purely animal, the body easing into an exquisitely sensuous scent-zone. It's the first time I get the urge to try a new mod directly on my skin in the midst of a session – I've always waited until the next morning to test with a rested nose. But this couldn't wait. Duende 25 was begging for flesh. So I rummaged through the drawers to find a pump, screwed it onto the phial and sprayed.

This is our first meeting since Bertrand's creative crisis. When I walked in an hour ago trailing the frigid December wind, I knew he'd come up with new mods even though I'd suggested we could just take stock. I was glad he'd resumed working but starting to wonder whether there was any point to my testing the mods he handed me after each session since he never waited for my remarks. So I asked him straight out: did anything I said matter?

'Until I'm happy with what I've done ...'

He waved his hand in front of his face and torso as though he were running it along a glass wall. Then backtracked. Of course, he listened: that was why he'd been in such a jam, trying to work my requests into the formula. But the Habanita accord I'd asked for was just too old-fashioned. In fact, he told me as we settled down to smell the new mods, he was still wondering whether the orange blossom and incense accord was such a good idea.

'You have to tame it, make it smile, lighten it up, and I'm not going to get there with incense! We'll end up with something *imbitable* and that's not the aim of the game.'

Imbitable, pronounced 'ahn-bee-tah-bluh', is a French slang word which means incomprehensible, unbearable. Bertrand's clients have been begging him to steer clear of his usual *imbitable*. He doesn't want to do *imbitable* any more, even for the sake of originality. This isn't the first time Bertrand mentions the need to be less quirky. But if he weren't experiencing a tug-of-war between composing on his terms and giving clients what they want, the subject wouldn't be cropping up so often, as though he were trying to convince himself.

Wait a minute … did he mention clients? We've never discussed the specifics of pitching Duende to one of the brands he works with, but that would be the logical final step of our creative venture and it's always been tacitly understood that we'd take it. I've been waiting for Bertrand to bring the matter up. The product probably needs to be closer to its final form before it's shown to anyone.

That time may not be far off. The proof is in the sigh.

I'm not staging it for Bertrand's sake. Right now, he's got his back turned to me. He's writing out the formula for Duende 28, also a first: he's never formulated right under my nose. He thought it up as we were discussing mods 25, 26 and 27, all variations on the idea of Habanita rather than a literal rendition. Habanita's coumarin has been swapped for tonka bean absolute, which naturally contains coumarin but gives off richer almond, hay and tobacco effects, along with slightly roasted notes. Bertrand has added cistus to N°26 and tobacco absolute to N°27, but N°25 is the one we're interested in. There's flesh in it, flower flesh: like sucking the nectar out of a plucked blossom, I blurt out. Bertrand says that's exactly what he was aiming for: orange blossom honey, nectar, petals … But the vanilla is a little overwhelming. Rather than taking some out, he decides to cover it up with the green notes of Duende 5. That's what he's doing now: one last mod before we wrap up for the day.

He's often teased me about being such a chatterbox, so this time I'm shutting up and letting him concentrate. Except, that is, for the sigh. Clearly, he hasn't heard it. He's scribbling away, mumbling about putting in a trace of this or that, reaching out for various materials ... So I go on sniffing my arm like a good girl.

Then it comes out. Not just a sigh. The Moan.

And there goes Bertrand, blending away obliviously. By the time I breathe out the second moan, I decide he should be apprised of the situation. After all, he's got a grown woman going loose-limbed in his lab just by sniffing his stuff. He really should be paying a little more attention. I clear my throat. Still no reaction.

'Ahem ... Bertrand?'

'Yup?' he answers, barely looking over his shoulder.

'I've ... uh ... been making a lot of little noises that should be pretty flattering to you.'

That catches his attention. He swivels on his stool to face me.

'What do you mean?'

As I launch into a mini-recital of the orgasmic noises I've been cooing, the Roadrunner's question pops into my mind, so I add:

'Of course, you know as well as I do that those sounds can be faked. In this case, though, there wouldn't be any point, would there? This came out spontaneously. I'd say it was a very good sign.'

He nods.

'Go ahead. Let loose ...'

Then he goes back to work. But we're both giggling.

30

'You should tighten your belt. You're flashing plumber's cleavage.'

Bertrand's iPhone is booming its 'Sonar' ringtone from someplace in the lab and he's been frantically trying to locate it, digging into his pockets, shifting sheaves of paper. He's just leaned down to rummage through his satchel.

'Excuse me?'

He straightens up, phone in hand. It's stopped booming.

'You know ... the bit plumbers bare when they've got their head under the sink?'

I point to the small of my back. He grins sheepishly and hoists up his jeans.

'Not a pretty sight, is it?'

Actually, wardrobe malfunction apart, Bertrand's quite spiffy today in a hip-geek way. He's swapped his rimless rectangular glasses for big cherry-red Buddy Holly frames, topped off with a narrow-brimmed grey tweed trilby. His long-sleeved tee-shirt is adorned with the face of a Papuan in primary-coloured tribal paint, who stares out from where nipples should be so that I never quite know who, of Bertrand or the Papuan, to look in the eye – it's the male version of Wonderbra's 'my eyes are up here.'

The iPhone bloops to indicate he's got a message. After listening to it, Bertrand pours out a string of expletives. I've always suspected he had a temper … While he launches into a series of calls to settle what seems to be a delivery problem, I go on serenely sniffing my wrist. This is our last session before Christmas and, this time, he hasn't presented me with a new version of Duende. We've simply been debriefing mod 28. I'm not sure it's because I complained he never waited for my input before tweaking: he's just been too busy to tackle number 29, I suspect. When I walked in half an hour ago, he was already grumbling '*merde*' as he hurriedly weighed a formula that a client suddenly wanted to get before 6 p.m.

Mercifully, he's not taking his stress out on me. As soon as he's settled his problem, he apologizes for the interruption and we resume our conversation in tones that feel much mellower than they've ever been. We've got good reason to feel mellow: we're both pretty happy with the direction Duende is taking, though Bertrand is wondering whether the green top notes aren't a little too cologne-like. I start prattling on about the fact that Latinos love colognes. When I stayed with friends in Spain I used to see giant bottles of *Agua de* this or that sitting in the bathroom for the whole family to splash themselves with. The effect can work itself into a Spanish story, I add.

'Still, I'll try to mask it a bit, make it more sophisticated. But it's important to keep the cologne effect, because at L'Artisan Parfumeur, they quite liked it.'

'Hold on … Why should you take into account what they like?'

Bertrand puts on his poker face.

'Well, I've been telling them about our project.'

'And?'

'They're interested.'

Shouldn't Bertrand be helping me off the floor at this point? Have I still got a pulse? Granted, I'm a sophisticated Parisian and squealing is beyond my moral reach, unless I'm in the general vicinity of any spider whose body is bigger than a raisin. Still, I'm taking this much too calmly. Is it because I knew all along this was bound to happen?

Duende and L'Artisan Parfumeur feel like a perfect fit. They've done a lot of travel-based scents, a few of which are Bertrand's. Though he's been taking on other clients, he's still their star perfumer, so they were the logical people to turn to.

But we haven't got a name yet! The earlier we find something that hasn't been copyrighted all over the planet, the better. I've known from the start Duende wouldn't be available. Before I gave Bertrand the book by Federico García Lorca, he'd called the project 'Séville Semaine Sainte'. Would that stick? Bertrand shakes his head. Too religious. We can play neither on 'Fleur d'oranger' as L'Artisan Parfumeur already has one, nor on 'Orange blossom', which exists in its sister brand Penhaligon's. 'Azahar', the Spanish word, has already been taken by the Spanish designer Adolfo Dominguez, though Bertrand quite likes the word 'Azar' which springs from the same root and means 'chance'. Maybe we could play on that. I'm groping round for words related to Holy Week …

'How about *madrugada*?'

'I love it! *Love* it!' Bertrand gushes.

'What does it make you think of when you hear it?'

'I was going to say it sounds very Spanish, but …'

'… that's comment degree zero!'

We burst out laughing.

'OK, so what is it, a dance?'

I explain that, in Spanish, it can either mean 'the wee hours' or 'dawn'. Pronounced with the lazy Sevillan accent, dragging the last two vowels together, *la madrugá* is also the period between midnight and noon on Good Friday – the very night I spent in Román's arms – when the exhausted crowds who've been partying all night in the wake of the religious processions blend in with the soap-scrubbed, cologne-splashed families rising for Mass.

As Bertrand writes an email to L'Artisan Parfumeur's main office to ask them to check up on the availability of the name, I start stuffing my notebook, phials and recorder into my hand-bag. When we'd set our appointment, he'd said he needed to be out by five thirty, and it's past six already. I'm just waiting for him to click on 'send' to take my leave.

But as soon as he does, he starts telling me about the way he'll tweak Duende. He wants to make it even more honeyed and nectar-sweet; he's been thinking about adding thyme instead of the lavender he used in the first two trials: it would bridge the cologne and incense effects. The latter he can amp up, but the material he'll be using won't express itself in the top notes. Fine, I can live with the incense note coming in later. After all, that's what happened in the story: waiting under the orange tree until the procession got nearer ...

As we sniff my wrist to follow the development of N°28, we agree the base notes are too static and un-contrasted. We'll work in more darkness, Bertrand says, but not the abrupt, mineral darkness we started out with in Duende 1 and 2. He wants to add tobacco to warm up the incense and play off the animalic waxy notes.

'We need to make it more mysterious ... We're not in the chapel yet, but we're getting close!'

I make a move for my coat. He grabs his too, puts on his little grey hat and says:

'So ... do you have time for a drink?'

I'm not due at my dinner party for another two hours. It turns out Bertrand can't go home before picking up a parcel near Bastille, and it won't be ready for a couple of hours either. So we head out into the night, try the nearby cocktail bar – no tables – and end up in an empty café on the rue de Rivoli in front of two steaming pots of frothy milk to pour into our hot chocolates. Neither of us is the 'have you seen a good movie lately' type but, as it happens, last night I saw Henri-Georges Clouzot's *The Mystery of Picasso*, which shows the master at work practically in real time. At each step, you want to say, 'Stop, this is it,' and then he paints it over, twists it and warps it until he chucks it all and says: 'Now that I know what I want, I'm starting all over again.' That feeling of seeing a piece of work morphing over and over again is what I'm getting from Duende.

'I don't know if Picasso knew exactly what he wanted,' Bertrand says, 'but I know that even if a piece seems to have integrity, if it isn't exactly what you want, you refuse it. You know exactly what you *don't* want, even when you don't know exactly what you want or where you're going.'

That's what's been happening with Duende, he explains. He's discovering it as he goes along, finding out what he can exploit from its current state, what will yield a new form. He'll pull on that form until it takes on a perfect volume. Sometimes he can achieve an extraordinary volume after one or two mods because the project's been well conceptualized beforehand, usually when the structure is simple and straightforward. Sometimes he has to start over countless times before finding exactly what it was he was looking for.

As Bertrand falls back on the metaphor of the labyrinth, I tell myself our conversations also follow labyrinthine paths: we often seem to pass by the same points but never quite at the same level. Of course, that must be the pattern of most conversations between people who meet regularly and are developing the kind of *culture-à-deux* that sometimes becomes the foundation of friendship …

So although we've been through this before, I bring the issue up again: isn't he taking on too many projects? He answers that, as a freelancer, he can't afford to turn down contracts. I shake my head. Someday, he'll get to the point where he'll be able to say no; where clients will say, 'We're hoping Duchaufour will make our new perfume, he hasn't agreed to yet but …' Bertrand says

'I hope so,' but he doesn't look convinced. I raise another issue: his style is so distinctive he can't help imposing it. What then of the brand identity of the various houses he works for? Not his problem, he answers. Clients ask for a product, he gives them the best he can, end of story. I shake my head again. I'm not buying his 'I'm just a guy making a living' line. Bertrand's one of the few perfumers I know who's got enough vision to art-direct himself. Even if he no longer wants to do *imbitable* stuff – and I'm not quite convinced he doesn't – he cares enough about the art not to be content churning out products, no matter how good they are. At that, he looks up:

'OK, right, so I've done a few juices that smell good while managing to be a bit original. But it's still classic perfumery. I want to move on to something completely different. Work on another language. Something more modern.'

I suddenly realize that, for the past year, I've been somehow waiting for him to say that. I was sure he would, sooner or later. As we pay up and head for the metro, he starts telling me about his new ideas. He goes on talking as we wait on the platform and shove our way onto the train. It's the rush hour three days before Christmas and we're jammed against each other in the crowd.

'You sure smell good!'

'Are you surprised?'

We're still talking as we are spat out by the crowd onto the platform of the Bastille station and wind our way through the piss-reeking corridors to the nearest exit. On the place de la Bastille, we move aside to avoid the hordes of last-minute shoppers, but it's too cold to stand and talk so the talk dies down. We exchange pecks on the cheeks and a hug and wish each other a happy holiday, and then he's zipping off to his appointment and I'm on my way to my dinner party, walking towards the place de la République along the boulevard Beaumarchais with its Harley Davidson dealership, camera and art supplies stores, weaving round icy puddles in my Italian kid-glove boots, fur coat flouncing about my legs. Next year's the year Duende may come true: I couldn't have found a more exciting gift under the tree.

Merry Christmas, Mr D.

31

Every time I go to Ottawa to see my parents, I check the level of the two perfume bottles on the glass shelf under the bathroom mirror. They never seem to go down much: in one year, my mother has used up about ten millilitres of Maharanih, an orange and amber scent by Parfums de Nicolaï I gave her after I found out she used her Jean Patou Sublime as a room spray. It seemed sacrilegeous to blast those precious drops into the air as a cover-up for more offensive effluvia. But though my mother admits she sneaks a spritz on her bra or on a Kleenex tucked under her pillow, room spray is her only official excuse for enjoying fragrance.

I've given her perfumes now and then, hoping the paternal ban would be lifted eventually. The limited edition Baccarat crystal bottle of Trésor I received from Lancôme back when I was a journalist was a miscalculation though. It is described as a 'hug-me' perfume – perfect for a mom – but my mother finds it too cloying. Trésor is now displayed next to her collection of Swarovski crystal animals. The plainer bottle of Eau d'Arpège, which seemed fitting when I picked it up at the duty-free shop because the scent had been composed for Jeanne Lanvin as a gift to her beloved daughter, ended up in the purgatory of the guest-room closet, alongside Bal à Versailles, Max Factor Green Apple and the Chloé I bought on my first trip to Paris.

After I've taken stock of these, I sit on the bed and stare at the flotsam of my pre-Paris life: my books, from teething to grad school; the stern-faced Spanish doll in crisp red taffeta flounces; the teddy bear we once drove one hundred miles to retrieve from a motel where we'd forgotten it, worn noseless and armless by a decade of little-girl love; pictures of my beaming face at every age. 'What will you do with all this when we're gone?' my mother keeps asking me, turning my stays into an anticipated mourning period as I survey each object I can't bear to let go.

I haven't been here twenty-four hours and, though I love my parents deeply, I'm already going stir-crazy. So I pull on my heavy boots, wind a scarf around my neck, slip on my fur coat and head out into the winter, trading one stifling atmosphere for another – at –10°C, breathing becomes slightly harder. There is nowhere to walk to that isn't a neon-lit box selling the same things you can buy anywhere in North America, so I pace up and down the driveway and scare off a few squirrels, black velvet ribbons undulating from the whiteness to the grey, naked tree trunks. I take my small decant of Avignon out of my pocket, spray some on my coat and bury my nose in my collar. When my father quit smoking, he complained that my clothes carried the stench of cigarettes over from Paris cafés. Rather than laundering the contents of my suitcase before each trip to Canada, I decided to fight smoke with smoke, hoping that the incense wouldn't register as perfume to my father's nose. Now incense has become the signifier of my Christmas holidays, not because of its religious connotations – I must have gone to Midnight Mass three times in my entire life – but because its mineral, dry, burnt quality is the answer to the blinding scentless nothingness of snow …

I've also brought along a decant of Annick Menardo's Patchouli 24. With its daisy chain of smoky, burnt, medicinal and leathery smells softened by the old-book sweetness of vanilla, Patchouli 24 epitomizes Menardo's style at its most radical. She composed it for a man who didn't wear perfume and asked her for something that didn't smell of it. She figured he could always spritz his father's old leather flight jacket with it. She had only herself and one man to please: fortunately for perfume lovers, it also pleased the New York-based Le Labo and no brand could have been more suited to it. In *Perfumes: The A-Z Guide*, Luca Turin compares it to the smell of a storage room in the Biology Department at Moscow State University. I concur: it reminds me of my father's lab. I'd actually told this to Menardo when we'd spoken on the phone. She herself studied chemistry, biochemistry and medicine before perfumery and, when she learned my father was a pharmacologist, she surmised that he had used cresol (whose tarry, burnt, medicinal facets can be found in single malt whisky and smoked tea) for liquid chromatography, a laboratory technique designed to separate mixtures in order to analyse them.

As we're sipping the Laphroaig I brought home for Christmas, I tell my father that the smell of his lab may have inspired some of my quirkier olfactory tastes; the whiff of the stables I got when coming to visit him conjures pleasant memories when I find it in a perfume. At this, he blurts out:

'*Parfums Ibérie, trois nuits à l'écurie!*'

'Ibérie perfumes, three nights in the stable.' Excuse me? My dad has just turned eighty, but he's nowhere near gaga: in fact, he's still working.

The memory just popped up for the first time in decades when I associated fragrance with horses. He explains that, when he was ten years old, he sent in for perfume minis that were advertised in a magazine in order to resell them. His commercial endeavour failed miserably, he says, and what's more, the cleaning lady kept crowing an annoying little ditty about 'Ibérie' perfumes smelling like the stables.

My Google queries yield no perfume or perfume house called 'Ibérie' or 'Iberia', but I do find a French house, Ybry, that developed its activities in North America in the 30s. I wonder what the horsy one smelled like. Narcissus? Leather? Was its smell what put my daddy off perfumes? Or was it the cleaning lady's jeers? It can't be the trauma of his failure as a door-to-door salesman: he put himself through his first year in college selling Fuller brushes ...

I'd have never imagined that my father had made an early foray into fragrance before becoming perfume-averse. Now I'm doing it the other way round, lifting the curse of Ybry by dreaming of an Iberian perfume. I've been meaning to show Duende to my parents, but although I've often spoken about it, they don't seem to be curious. It's only once my suitcase is packed for my return trip that I ask my mother whether she wants to smell it.

'Oh. It's strong' is all she'll volunteer, with a tentative smile.

An hour later, when I join her in the kitchen, I hold out my wrist again. She agrees it's softer now. She likes it better.

As for my father, I've been wondering whether perfume still bothers him as much as it used to: he's never made a comment on the fragrances that must permeate my clothes even when I don't apply any. I ask him whether he *does* smell them.

'Yes, in fact, I do,' he answers with a sly little smile.

Somehow I feel that if he hasn't made a comment, it's for the same reason he never uttered a word against my husband until I announced my intention of divorcing him: out of respect for my choices. Still, I press on. Does he want to smell Duende? My mother looks as though she's about to catch my wrist when I raise it to my father's nose. Her expectant, amused, slightly defiant gaze flits from my face to his.

'It smells ... like perfume,' he finally says with a baffled chuckle.

'But why perfume?' my mother asks me. 'I don't understand. You've always liked it, but there were many other things you liked you could have chosen to write about.'

Why indeed?

I could answer that each tiny puff of beauty is my stand against the ever-more-standardized world of shopping malls and big-box stores where I've chosen not to live; the essences of flowers, spices and woods grown in warmer climes a protest against the four-month-long winters of my youth. I could say that I drench myself in sweet scents for all the times my mom dabbed on a drop of Bal à Versailles in secret or sneaked a spritz of Sublime on a Kleenex, ever the daughter rebelling against the Law of the Father. I could say that perfume is the only thing that allows me to breathe when the cold air catches in my throat and I'm smothered in childhood memories; that perfume is the language of my chosen country; that in many ways it *is* my chosen country, invisible and borderless: that's why I need to learn its language.

I don't. Instead, I slip my phial of Duende 28 into the regulation Ziploc bag next to my Ventolin, lipstick and mascara before we head for the airport. If it comes to it at the security checkpoint, I'll dump the Ventolin.

32

When Bertrand and I came to this café before Christmas, we'd had the whole banquette to ourselves. Now we're stuck between the window and the railing of the staircase leading down to the cellar, an icy draft rides up my skirt each time the door opens and a dozen people are howling over the screeching coffee machine. But Bertrand's got a new trainee and there's no space for the three of us in the lab, so we'll have to make do with this set-up.

Nine phials are lined up on our tiny round table. Guess the new trainee is helping. It's been a month since we last met. I came back from Canada with a bad case of flu and stayed curled up in a foetal position in a puddle of sweat for a week. When I emerged, as I had had no news from Bertrand, I figured he was swamped with work and gave him a breather; I'd wait and see if he contacted me of his own accord. I broke down first – there was no earthly reason why I should be following the wait-for-the-guy-to-call rule with him – and found out *he* thought I'd been giving him the silent treatment. So he'd gone ahead and showed the latest mods to the team from L'Artisan Parfumeur. He only told me about the meeting after the fact. They loved it and they're considering launching it next year.

Well, Mr D., I'm not popping the cork off the bottle of Moët yet. I'm delighted the team loved it. But, for the time being, the only thing that matters to me is what *we* want Duende to become, I tell Bertrand once we've settled in front of our hot chocolates. I don't want anyone else interfering with the process. Not even the client.

Bertrand seems a little taken aback by my reaction. I'm surprised myself at how territorial I'm feeling, though to be perfectly honest I'm probably just irked at having been kept out of the loop – but where's my place in that loop? 'Muse' isn't a position on organizational charts. He explains that it was just a quick meeting and that he only showed them two mods, the one I'd left with before Christmas and the latest one, but none of the intermediary versions. There are things he'll want to go back to.

'You'll see. *You* tell me.'

Now *that* was a deft bit of defusing … So while Bertrand finishes numbering the blotters, I tell him how I made my parents smell Duende on the day I left Canada. He asks me if I've ever figured out why my father was so intolerant of perfume.

'Maybe he's really hyperosmic, who knows? Or maybe he's just a control freak.'

'I guess a lot of old-school scientists were like that,' Bertrand muses. 'Like my dad. Everything had to be controlled; everything had to be rational. Now that I think of it, I'd never realized this before but my father never wore fragrance either.'

Bertrand's father was a pedologist. At first I think it's got to do with feet, but no, that's podology. Pedology is a branch of geology, the science of soils. The late Dr Duchaufour was one of the leading authorities in the field.

Swapping childhood memories always does the trick, doesn't it? Once Bertrand and I have performed our bonding ritual, we can resume plotting the progress of Duende. There aren't nine mods but seven: N°29 is just a base and N°34 has gone AWOL. All are tweaks on N°28 and running through them is like watching the scent morph as the different materials twist, blow up or tamp down its accords.

In mods 30 and 31, the incense note is produced by new materials (incense absolute instead of the essential oil, and the synthetic molecule cedramber); mods 32 and 33 have an added tobacco note produced by cistus. Bertrand hasn't smelled them for three weeks and he's pleasantly surprised by the way they've evolved after macerating. Like all resinous materials, cistus has the effect of a lacquer, smoothing out and polishing the other notes. This is where we wanted to go: darker, deeper inside the chapel.

Mod 35: the diesel fumes of rose oxide suck the life out of the orange blossom. Interesting, but not the story. Scrap it.

Mod 36: powder puff. Musk, trailing its usual partner-in-crime: vanilla. He quite likes it. I don't. It smells too old-fashioned and, besides, I tell him, he should be giving musk and vanilla a break. He's been using them a lot lately. Scrap it.

Mod 37: 'Black pepper, pink pepper, incense: boom, boom!' Bertrand says about his new take on the top notes. The two peppers pull the incense upwards. The pink one adds sparkle with a natural feel, brightness without resorting to zesty notes. But even though there's only one per cent of the synthetic musk muscenone in the formula, we both find it annoyingly invasive. 'That baby-skin effect bugs me,' Bertrand grumbles.

We settle on a combination of 37 for its more incensey top notes and 33 for its darker base. Duende shouldn't be *too* dark but, still, a lot more should be happening in the base notes. Whatever became of the African Stone, for instance? No problem, Bertrand says, we can work it back in. There's a rusty, old-penny facet that'll go towards conjuring the blood note.

'And then there's also that old church incense formula you brought me ... I keep going back to it. It's a perfume in itself!'

In early December, when Bertrand was struggling with the incense note, I'd done an internet search on Catholic incense and come up with a formula from 1834, a blend of several resins and balsams, sandalwood, cloves, cardamom and, surprisingly, lavender.

I'm really getting the hang of this muse business, aren't I? Still, we aren't there yet. At the rate we're going, and taking into account Bertrand's punishing workload, we agree Duende won't reach its final form for another two months at the earliest.

So: February … March … April. 22 April. One year to the day after our first session. In 2011, 22 April falls on Good Friday. Duende must be done on time for the Madrugada.

'What if I took you to Seville for Easter, then?'

I've been updating Monsieur on Duende's progress. We pause to breathe in the briny, nutty aroma of our steamed scallops in seaweed. The chef of Le 21, Paul Minchelli, is a temperamental Corsican and you can't get a table at his hole-in-the-wall restaurant on the rue Mazarine unless you know the secret handshake. Several major perfumers do – Jacques Polge and Jean-Claude Ellena are clients – as well as art dealers, writers, movie stars and film directors. And Monsieur.

'I'd love to! There couldn't be a better way to end the story.'

As soon as I've said the words I wonder which story I had in mind.

Seville was what drew Monsieur to me in the first place twelve years ago: I'd just published an erotic short story set there and its contents were discussed at the dinner party where we met at a mutual friend's house. After that, I didn't hear back from Monsieur for six months and then, all of a sudden, he wanted to meet me to discuss the story, which he'd finally got round to reading. Then I met Monsieur all over France and Europe but we never made it to Seville. Going there to measure up Duende against its original inspiration sounds like a plan.

What will happen after that, supposing Monsieur really does take me to Seville, which I'll only believe once I'm aboard the airplane? I can't see past the moment when Bertrand and I will say, Stop, we're done, the perfume has become what it asked to become … For a year, what's been driving me is the thrill of meandering in the labyrinth. Now that L'Artisan Parfumeur has peered down into it, I realize I'll soon be airlifted out. Duende will no longer be the thread of my journey. Seville's heart-rending beauty could be the knife I use to cut loose.

But why bother to go to Seville? Won't Duende take me there instantly once it's done?

Short answer: no. In fact, the next time someone brings up the perfume-as-instant-flashback cliché, I may scream. Along with asking me whether I've read Patrick Süskind's *Perfume* or seen the film, it's the first thing anyone mentions when I say I write about fragrance. How it brings back memories. And then, unfailingly, the most overworked piece of pastry in the history of literature splashes into the teacup: 'I raised to my lips a spoonful of the tea in which I had soaked a morsel of the cake. No sooner had the warm liquid, and the crumbs with it, touched my palate than a shudder ran through my whole body, and I stopped, intent upon the extraordinary changes that were taking place …'

In France, any object that carries you back to your past is dubbed *ma madeleine de Proust*, 'my Proustian madeleine': it is probably the only thing most people know about the seven-volume *In Search of Lost Time*. So when the word 'madeleine' comes up in a conversation about fragrance, it is some vague notion of the Proustian epiphany that is referenced rather than the patron saint of perfumers (the small scallop-shaped sponge cake was actually named after the maid of the Marquise Perrotin de Baumont, Madeleine Paulmier, who invented it in 1755). The madeleine, still commonly served at breakfast or tea, has transmogrified into a master key to involuntary memories.

But is the 'madeleine effect' verifiable? Are olfactory memories really more emotion-laden, more indelible, more immediate and more accurate than those conjured by the other senses, as most people seem to believe? In *What the Nose Knows*, the sensory psychologist Avery Gilbert sums up four decades of research into odour-evoked memories and concludes they aren't: 'rates of forgetting [are] the same as for sights and sounds'. Memories of odours are 'subject to fading, distortion and misinterpretation', just like any other. In that respect, smell has no special status. According to Dr Gilbert, the element of surprise may be what makes olfactory memories so striking: 'Because odor memories accumulate automatically, outside of awareness, they cover their own tracks. We don't remember remembering them. The sense of wonder that comes with the experience is, like all magic, an illusion based on misdirection.' In *Éloge de l'odorat* ('In Praise of the sense of smell'), Dr André Holley, a neuroscientist at the Université Claude Bernard in Lyons, puts it another way: what's so striking is 'the contrast between the immateriality of the cause and the emotional strength of the effect'. Whereas the 'visual creatures we are' can summon images from the past more readily, the encounter with a smell is 'rare, more unexpected and therefore more precious'.

I can't deny the evocative powers of smells, and I won't. It may well be that my relationship to perfume has been perverted by my approach to it, or that after years spent alongside Parisian intellectuals I've caught their bent for taking the opposite view of anything that sounds like received wisdom. The fact is that the cliché of perfume-as-*petite-madeleine* riles me because it demotes its object to the rank of smell and reduces it to being the instrument of an emotional experience. What has been bypassed is intelligence: the madeleine school of scent appreciation isn't meant to make you think, but *feel*, which costs a lot less effort. But the perception of perfume is not a purely emotional experience. Fragrance is a cultural, artistic construct, with its own rules and history. As a product caught up in history, it adds new scents to the catalogue of memories; as an artistic product, it *plays* on cultural memories and connotations as well as on the personal experiences of the perfumer or the wearer. 'Wouldn't a radically new smell be threatened with a protracted purgatory in the register of bad smells, because it's not in accordance with memorized sensations?' writes Dr Holley. 'It is only through a subtle association of the known and the novel that new olfactory forms can be adopted.'

Memories are the ingredients of perfume-making, but the end result shifts them, inserts them within a new form, much as the madeleine leads not back to Marcel Proust's past but *ahead*, towards *In Search of Lost Time*: 'Seek? More than that: create. It is face to face with something which does not so far exist, to which it alone can give reality and substance, which it alone can bring into the light of day.'

Proust could have stayed at home spreading cake crumbs on his bed while mawkishly reminiscing about his childhood. Instead, he fished out his soggy madeleine and went on to raise 'the vast structure of recollection'. It was his future as a writer he was seeing in that teacup.

I'm not asking that much from Duende and, besides, my experience of orange blossom and incense is the very opposite of the Proustian flashback: I was *aware* of smelling them at the time and I committed them to memory. I can summon the smells mentally and identify them in a fragrance. Perhaps if someone captured the headspace of the Semana Santa, re-created it and plunged me into it unawares, I'd experience the same uncanny déjà-smelled feeling as *le petit Marcel* and flail helplessly, caught in a temporal short-circuit, as I try to identify the source of my emotion ... In fact, it *did* happen to me once, with the orange trees in Marrakech.

But I know full well that my perpetually mutating chimera won't actually smell of the Semana Santa or carry me back there. I don't expect it to. I don't even *want* it to. Just as the story I told Bertrand last year is a reinvention so vivid it has crushed my memory of what truly happened on that night in Seville, Duende is not a travel memoir but a piece of fiction writing itself as we go along. It isn't taking me back in time. It is moving me forward.

Duende is the future memory of what I'm living *now*.

33

'Yes … Oh yes!'
This comes out as soon as Bertrand thrusts his blotters under my nose. He's gloating too, eyes flashing behind his hipster glasses.

'Oh, this is becoming good, it's becoming so gooooooooooooood! I'm finally getting where I wanted to go.'

When he emailed me two days ago saying we absolutely had to see each other before he left for his skiing holiday, even for half an hour, he sounded excited (me, not so much – half an hour?). It turns out he had every reason to, and my spontaneous 'Yes' is even more of a vindication than the Moan. It's as though a veil had been ripped off Duende. It's come into sharper focus, with stronger contrasts, its burnished gold whorls and flesh-white blossoms catching the brass-bright cool moonlight or the first dawn sun … There's a technical reason for this. Bertrand has removed cedramber and rose oxide because he hated, just *hated*, their 'empty aspirin tube' powderiness (he's cute when he gets so worked up about a note: it's all he can do not to stamp his foot and spit on the floor).

But he's also added something to Duende that's made it come full circle from the very first proposals to mods 53 to 55 (he's been working like the devil: the last time I saw him we'd left it at 37). When he tells me what it is, my heart sinks.

Lavender.

To me, the stuff smells like those tourist souvenir shops in Provence with the piped-in chirp of cicadas, the ones that sell mustard-coloured table cloths with olive patterns and tulle sachets stuffed with dried lavender flowers that always end up reeking of dust. It's the smell of the colognes Anglophile Frenchmen splash on to match their Prince of Wales-patterned suits. It's the noxious fume of Brut-doused louts. Every time I test a lavender fragrance, I feel like I'm sprouting a chest rug and a glass of Ricard.

So why in the name of Mary Magdalene is lavender so *right* for Duende all of a sudden? How did Bertrand figure out it was just what was needed to topple it over into the Yes zone?

Is it my fault? I'm the one who brought him the 19th-century church incense formula that featured lavender. I'm also the one who reminded him in an email that, before Christmas, he was thinking of adding thyme, which he'd completely forgotten about. He was dickering with the thyme and getting nowhere. From there, he segued into another aromatic note, lavender: it was his idea from the outset and he's nothing if not a headstrong Capricornian. So he put in Provence lavender, but also a special type of absolute from Spain called the Luisieri.

The minute he makes me smell it, the Luisieri blasts away any preconceived ideas I had about the stuff. It's a perfume in itself, with a smooth, almost liquorice-like burnt resinous cistus darkness, dried fig, amber, tobacco, honey and animalic facets, along with the combustible aroma of the maquis. You can even see how it could be tugged into an incense note. This is so tough and tender that, if it were a man, I'd date it. And it fits perfectly into my original story since it is also known as Seville lavender.

Bertrand and I once talked about the common point between his conception of perfumery, his practice of T'ai Chi and his love for African art, and I suggested it was all about the *justesse du geste*, the rightness/sureness/accuracy of gesture. You could also call it *elegance*, in the scientific sense of the term. Lavender is an elegant solution.

But lavender in Duende is also right in a way that reaches beyond its necessity in the formula, or Bertrand's wish to reference Spanish fougères and reinforce the Habanita effect with the cistus-like facets of the Luisieri. The newly unveiled Duende is reaching into *me*.

After leaving the lab, I cross the Pont des Arts from the Louvre to Saint-Germain-des-Prés. I've lived so long in Paris that each walk has become an increasingly poignant, unwilling pilgrimage, moving in and out of ghost-ridden zones as memories seize me and slap me and send me spinning into all of my pasts … Lingering in front of a bookshop, absent-mindedly sniffing my wrist, I remember I stopped here once before, making time before meeting Monsieur at a hotel on the rue des Saints-Pères early on in our affair, and that's when I suddenly understand why Duende feels as exhilarating as falling in love.

To me, lavender *was* the scent of falling in love: the one note Monsieur insisted on when he asked me to pick out a new fragrance for him. It reminds me of the Mouchoir de Monsieur he wore for years, so it's almost as though Monsieur had invited himself into my story of Seville a decade ahead of the time we met. And as though he'd slipped into Duende, ten years after I bought a candy-striped knitted cashmere scarf identical to his, and swapped them secretly so I could smell Monsieur on me when I went back home to my husband.

Should I recant on my rant about perfume-as-memory? Bertrand's intuition has bounced off the curves and hollows of my soul like an ultrasound scanner to sculpt the volumes of Duende from the spectres he's found there.

But it's not just the lavender. It's Duende itself. It is stirring from its chrysalis to become its own creature and starting to play games with me. For the first time, I crave smelling it, but I sneaked sniffs of N°54 and N°55 for a week before actually wearing it, as though the waiting would make it even better. Then, after one week's wait, I just grabbed N°55 and sprayed it on the way you lean in for that first kiss – offhandedly, no lead-in, no build-up, so that it's done with. But instead of going into raptures, I found myself kicking into analytical mode almost unconsciously – a background programme running as I went about my daily tasks, as it does when I'm wearing a scent to review it. I'm analysing Duende like a finished product. And that's new too.

Hi Bertrand,

I don't know if you've worked on the next mods yet but here are my impressions of Duende.

1/ I adore it. It moves me. There, I've said it.

2/ I get spontaneous compliments on it. One of my readers, who I met to give her a scent she'd won in a draw, wrote this to me one hour later: 'I forgot to ask you what perfume you were wearing. It's beautiful, it even borders on the sumptuous. And it's really creative. I've never smelled anything like it.' So there you go! But …

3/ It's behaving a bit weirdly. After that first burst of lavender which really lights up the blend – that was a genius idea – the honeyed tobacco and balsam stick close to the skin, just a smudge I can barely perceive. But when I leave a closed room where I've been for a while, I come back to find an entirely different smell, all incense and wood. It seems to be living in two different spaces, one too near, one too far, both just out of my grasp … The two volumes need to be bridged, what's tamped down pulled upwards and outwards, the lavender thrust deeper into the heart. What I smell on my skin is gorgeous, but it needs more volume!

D.

Hi Denyse,

Wow! Thank you, Miss, for this thorough update. I'm very happy to read what you tell me. It really confirms the ideas I have about the product. We're closing in on it and I think I know exactly how to finish it. I'll drop you a line when I've got something new to show you.

B.

34

We're almost there. The orange blossom is back. No longer the odd-smelling raw material with hints of snapped peapods and raindrops on hot asphalt I smelled that first day in Bertrand's lab, the orange blossom absolute that tugged at my memory. Now it's fully fleshed out, a flower, a tree, a plazaful of trees with their clusters of white stars and their dark waxy-green leaves and the wrinkled unpicked bitter oranges from the previous year, as though lighting up the night-time scene with lavender had released the flowers' scent – Bertrand had told me at the very outset that we needed to find the nightlights, the flares …

Everything he's been trying to put into Duende all along is finding its place at last. This is the nectar-laden floral accord he'd been perfecting when I derailed the process by bringing in Habanita. But now it's got authority. It *belongs*. He nods.

'That's it, exactly. When I first used the lavender, I was trying to close the loop before I'd even drawn the circle. Now we've got our structure, so I can put back everything I had to take out, the way it's supposed to be, where it's supposed to go.'

We settle on N°63 after smelling mods numbers 56 to 64: it's the one where the orange blossom expresses itself most beautifully against the backdrop of the balsamic resinous base. Each element in the formula adds its vibrancy to the others: the wood accord boosts the slightly spicy facets of the jasmine; the jasmine, boosted by the wood, resonates with the orange blossom through their common indolic facets – Bertrand says it's 'like a domino effect'. As a result, all the secondary accords, the green, the balsamic, the musky, have also become more ample without crowding each other.

I'm also getting … something boozy? Bertrand nods again, looking impish.

'I've worked in an ethereal, alcoholic effect in the top notes.'

'Because …?'

'Because it's a boozy night, you told me so yourself!'

True. We *were* downing quite a lot of manzanilla. Its almondy, woody aromas are easily conjured by the notes in Duende; the tiny drop of blood that seeps through carries the wine's saline sea breeze … The rum absolute he put in doesn't read like rum, but as it whooshes upwards it gives an exhilarating jolt to the green top notes: makes them smile. That's something he'd been looking for too.

But as everything in Duende is falling into place, my own position is shifting. Bertrand tells me how impressed he was with my report on Duende N°55:

'You really know how to smell now. I won't touch N°63 until you tell me what you think.'

It's taken me a year to get there; it's taken him a year to listen. Hasn't anything I've said up to now made any difference?

'You brought me the story.'

Ah.

'I had to wait until *I* was satisfied with what I'd done before I could listen …'

'And now you're satisfied. So what do you need me for?'

'Now I need you as an evaluator.'

That pinches a little. However useful an evaluator's input is, it's mostly technical and commercial: she's not working on *her* story but on a client's brief. It's not a moment of her life that's being bottled. Duende comes from me as much as it comes from Bertrand, and whatever I've been to him throughout the year, I *am* much more than an evaluator. Aren't I?

'OK, so you *are* a muse …'

It's the first time he's said the word. I'd kiss him smack on the top of the head, if he weren't saying it in such a teasing tone.

'… *and* an evaluator.'

OK, so now I could bang his head against the counter. But of course I'll do it. Who knows Duende better than I do, after him? I'm pretty sure that of all the people he works with, no one follows the development of their product as closely as I do. Do they?

'Never.'

I think back to what Christian Astuguevieille told me: that much of his work with perfumers was learning to talk with them. Those long hours Bertrand and I spent together, huddled over blotters, were spent looking for a common language. Bertrand corrects me:

'Looking for a common *emotion*, for the moment when we'd both go "ah" … A meeting of the souls.'

And that's happened during the last session. In a way, Duende is done. From now on he'll focus on rebalancing proportions without adding any new materials to the formula. My mission will be to keep track of certain aspects of the development that might need polishing: whether the cat pee facet in the blackcurrant base sticks out too much; if the aldehydes are too screechy; if the new ethereal, boozy effect works well. Whether the scent evolves harmoniously rather than splitting up into two different volumes like it did in N°55. How long it lasts on skin; if it carries; if I get compliments on it … But even if the next mods are just tweaks, we're well on for going over 70. How does that measure up to the other work he's been doing, I wonder?

'The most mods I've done were for a Penhaligon's product. I was up to 78. So I'd say it's getting to be pretty sophisticated.'

The competitive streak in me is speaking up and I make a mental note of pushing to 80. Suddenly Bertrand looks me in the eye, scowling:

'But, believe me, that's *nothing* compared to the way I used to work with my colleagues in the big companies! Those people who called the shots, they didn't have a clue about what they were doing. They kept changing their minds. We did *thousands* of mods, and there could have been ten masterpieces among them, but those were never the ones they picked! The perfumers work themselves sick on a chunk of the road that's two metres wide, when there's a whole highway of creativity that should be open to them! But no! No one ever goes past those two metres because, beyond, it's the unknown!'

Clearly, the memory of the years he spent slogging over commercial products still rankles. I've seldom seen him so worked up. But surely those days are over for him, aren't they? He shakes his head.

'No, I'm still restraining myself a lot, because I'd love to work on raw materials that are *imbitable*, but people wouldn't understand. They just wouldn't.'

There's that word again. I *did* suspect he wasn't being entirely in good faith when he said he wanted to move away from the difficult notes.

'But I'll do it anyway. We're going to try to impose a certain idea of perfumery. We'll tilt against windmills. I'm Don Quixote!'

Well, Mr D., don't expect me to play Sancho Panza, the paunchy peasant who trails after the deluded knight errant as his squire. But though Duende is far from *imbitable* – it's got the lushness of an old-time classic, like a modernist rewrite of a Guerlain – I'm more than willing to gallop alongside you as a fellow knight with a battery of atomizers tucked in my ammo belt. We'll spray those windmills until they spread the waft.

35

'I want to make the most beautiful oriental perfume possible,' Bertrand writes in an email, and suddenly I realize he's right. Duende is so much its own creature to me that I've never thought of fitting it into a slot, but the soft, balsamic accords and resins that form its base are indeed the core of the oriental, also known as the amber fragrance family. How in the world did that happen? Orientals are not my favourite genre. My inclination runs to the more tightly corseted, sensuous yet intellectual chypres. Had I been an oligarch's bit of arm candy commissioning a bespoke perfume, I wouldn't even have requested the notes found in Duende. But if Duende is an oriental, it is because Seville *dictated* it.

It was as one of the jewels of the Moslem world that Ishbilya, as it was then known, shone most brightly in the Middle Ages, as brilliant and refined a capital as Baghdad and Damascus. Its great cathedral, the Giralda, rose around a minaret whose twins still stand in Marrakech and Rabat; the cathedral's patio of orange trees was once the courtyard where the faithful performed their ritual ablutions. To this day, from the jasmine-choked alleys of the old Jewish quarter to the fountains glimpsed behind the wrought-iron grilles of its traditional *casas de patio*, Seville still winds its way into the world of the Arabian Nights. And yet this gateway to the East became the door to the West: the gold splashed on the altars of its Renaissance and Baroque churches was ripped away from the entrails of the Spanish Americas; the treasures of the New World that passed through Seville until the late 16th century turned it briefly into one of the wealthiest cities in Europe.

Seville's dual nature – looking to the East as one of the west-ernmost outposts of the Moslem world, then to the West as the treasure chest of the Spanish colonies – is reflected in the very notes of Duende. The incense comes from Arabia. The orange tree first grew in Asia. Lavender and cistus labdanum are Mediterranean. But vanilla and tonka bean originated in the Americas. Of course, you could say that almost every perfume is a blend of essences that found their way into the perfumers' palette after travelling the trade routes of the planet ... Or, conversely, that perfume is oriental by its very essence, no matter what you put into it.

Sophisticated extraction methods were already developed as far back as Ancient Egypt, but it was the Arabs who revolution-ized perfumery with the distillation of alcohol and rose water. Their techniques were then imported into the West along with aromatic materials through the Crusades and the merchant port of Venice.

But the modern perfume family called 'oriental' has little to do with the blends of Egypt, the Levant or the Middle East. Though it may be reminiscent of the sweet musky scents of the perfumers' souks, it is a Western reinvention, in the same way that Western artists, craftsmen and poets reinvented the Orient as precious goods were imported, travellers brought back tales and literary works such as *The Thousand and One Nights* were translated. Harems inspired fevered dreams of unfettered sexual-ity, as well as being a pretext for painters to depict voluptuous nudes. Writers like Chateaubriand, Lamartine, Nerval, Flaubert and Gautier turned the trip to the Orient into a literary genre. Inevitably, sultanas, odalisques and slave girls wound their way from paintings, poems, plays and operas onto perfume bottle labels. The fragrances were only vaguely related to Eastern blends through the resins and balsams that had featured in perfumers' palettes for centuries, such as benzoin, whose soft, powdery, spicy smell is partly owed to vanillin. Vanilla itself had been in use since the time of Louis XIV; after vanillin was first synthesized from clove oil in 1874, it found its way into perfume formulas because it was stronger and cheaper than vanilla tincture (the pods soaked in alcohol), though still a high-end material. Matched with cistus labdanum and bergamot, it formed what is known as 'amber'.

The oriental note as we know it seems to have originated in the mid-19th century in a perfume called Opoponax, by the British perfumer Septimus Piesse. But it was Aimé Guerlain who perfected it after the discovery of vanillin allowed him to overdose the note in Jicky, which hovers between the oriental and the fougère. The amber accord was later 'extracted' and sold as a base by raw material suppliers. François Coty mined this rich olfactory seam with his 1905 Ambre Antique, whose bottle referenced antiquity and whose fragrance, built around a base called Ambréine Samuelson, echoed the blends of Ancient Rome with their rich amber, resins and cinnamon. Coty would go on to invent two templates of the oriental family with the 1905 L'Origan, the ancestor of the spicy floral orientals later epitomized by Opium (the first product actually to be called 'oriental', according to the historian Elisabeth de Feydeau), and the 1921 L'Émeraude, which also featured a sweet oriental accord but with stronger aromatic and hay-like effects reminiscent of the fougère family. But though their early boxes displayed Persian-style motifs, none of the Cotys played overtly on oriental connotations.

It was not the Napoleon of Perfume but the Pasha of Paris, Paul Poiret, who capitalized on the Orientalist wave that washed over pre-World War I Paris in the wake of the Ballets Russes (Rimski-Korsakov's *Scheherazade* debuted in Paris in 1910) and of a new, unexpurgated and therefore vividly erotic translation of *The Thousand and One Nights* by the Egyptian-born Dr Jean-Charles Mardrus. Poiret had turned chic Parisiennes into sultanas with his harem pants and turbans; in 1912, he hosted an extravagant *One Thousand and Second Night* costume party staged with the help of Dr Mardrus. Unlike Coty's, Poiret's fragrances did explicitly refer to the Orient, with names like Minaret, Aladin or Nuit de Chine. And, according to Octavian Coifan, it was Poiret's Orientalist vision rather than his use of the traditional materials of oriental perfumery that led us to associate certain fragrance types with the Orient, most notably the carnation, violet and sweet notes already found in Coty's 1905 L'Origan, which Poiret reprised in the 1912 Chez Poiret.

But it is Shalimar, launched by Guerlain at the 1925 *Exposition internationale des arts décoratifs et industriels*, which would give its name to Art Deco, that is identified as the first oriental in modern perfume history. The reason is simple. Like Chanel N°5, it is one of the few continuously produced fragrances of the pioneering years. Like N°5, it is still widely sold and supported by lavish advertising campaigns. So in the same way that N°5 is claimed to be the first abstract perfume or the first perfume to use aldehydes, because the other contenders to the title are all but forgotten except by a handful of experts, Shalimar can claim to be the first oriental. But just as Chanel N°5 was, if not the first abstract fragrance, the first to display a consistent identity as such (the name, the bottle, the box, the notes), Shalimar was the first perfectly accomplished example of the oriental style. It had the notes. It had the name, inspired by the 17th-century Shalimar ('abode of love') gardens in Lahore. It had the romantic back-story: Shah Jahan, the son of Shah Jahangir, who laid out the gardens, was none other than the builder of the Taj Mahal, erected in memory of his favourite wife, Mumtaz Mahal. It had a correspondingly exotic ad featuring a veiled Indian woman, though other ads designed for the Anglo-Saxon market played both the oriental and Parisian cards by showing the Champs-Élysées: 'The chic, the verve that is Paris ... The mysterious, compelling allure that is the Orient ... The inspired admixture of both ... That is Shalimar.'

Though the Shalimar talcum powder I bought as a teenager is still tucked away in the closet of my bedroom in my parents' flat, I was never a Shalimar girl. Perhaps because, unlike its elder sister Mitsouko, it is almost too easy to love with its powdery notes that are a molecule away from veering into the edible. The boudoir where Shalimar flaunts her cleavage is clearly built over a patisserie: trust the French to segue from one kind of carnal appetite to another, turn them both into an art form, and bottle the result. As far as orientals go, Duende is actually a little nearer to the pastry platters of the Islamic world with its orange blossom, almondy tonka bean and honeyed beeswax.

An Iranian friend who once asked me to help her find a new perfume added: 'But not too oriental. With my type, that would be overkill.' There is nothing particularly exotic about my own physique but I *am* inclined towards Mediterranean plushness and I sometimes wonder whether that wasn't what drove me eastwards, where whatever charms I possessed would be appreciated. Going from Montreal to Paris was a logical, cultural leap over the ocean. But I had to come to Seville to start appreciating a body that had never consented to conform to contemporary Western norms. It is in Seville that my sensuous world pivoted on its own axis, just like the world itself had pivoted from East to West in Andalusia in 1492, when the last Moorish stronghold, Granada, fell into the hands of the Catholic kings the very year Christopher Columbus was claiming the Western Indies for them. Duende is an echo of my journey. The alchemical blending of the East and West, of oriental ancestry and French art, of materials from the Old and the New World ... The quintessence of my quest for the place where my body and soul belong. Southwards. Eastwards.

'Would you be willing to fly to Beirut tomorrow?'

The call came five years ago. I'd been warned by the business consultant who'd introduced us in Paris that Habibi wanted to use my services as a ghost-writer. When he phoned a few days later, I was invited – nay, *summoned* – to Beirut, rather like a scriptural call-girl. The details were arranged by a male secretary: I was to be flown business class, picked up by a chauffeur and accommodated in a boutique hotel on the Corniche. I accepted because I was, as often in my freelancing life, pressed for money, but also because it seemed very Graham Greene to be flown in on such a secret mission.

Habibi picked me up in the lobby at eight, a tall man in his late forties with the lazy, imperious manner of a first-born son and garter-belt-snapping eyes. In my standard-issue Parisian black I felt like a sparrow caught in a flight of birds of paradise among the designer-garbed Middle Eastern female clientele, but Habibi claimed he was glad to be seen with a woman who still had the nose God had designed her with: he was bored with what Beirut had to offer. In fact, I came to understand as we chatted over dinner, Habibi was bored, period, since he'd sold off his business. I waited for him to tell me what job he'd asked me over for, but clearly this was a matter that would be dealt with later on.

'Do you have a lover at the moment?' he suddenly asked me.

Oh. So it was *that* kind of bored. I didn't make much of a fuss, answering that there had been someone, but that our affair had petered out because Monsieur was so much busier than when we'd met that he couldn't properly attend to me.

'So it stopped by suspension of payments,' Habibi purred.

I chuckled. He wasn't wrong, though his interpretation was, perhaps typically for a Lebanese businessman, couched in the terms of a financial transaction.

It was past midnight when we drove off into the April drizzle, and the car wandered in the eerily new city centre rebuilt on the model of French colonial-era architecture, pastel buildings encrusted with the glittering windows of Gucci, Prada and Burberry shops, a Disneyworld for label-shoppers edged with gutted buildings and the ruins of apartment blocks abandoned before they were completed at the outset of the civilian war. Beirut's lopsided smile, with its pristine implants alongside decayed stumps, was oddly compelling.

We did eventually broach the purpose of my trip over whiskies in the bar of my hotel. There would be a couple of people to meet who might be interested in writing their memoirs. I took an option on a vanity book for a lady friend of Habibi's, and then Habibi took an option on me.

As he never got up before noon, he let me have the chauffeur for the morning. But the chauffeur spoke neither English nor French and I struggled with what I thought was the Arabic for perfume, *attar*. The chauffeur's eyes in the rear-view mirror were puzzled until I resorted to miming dabbing my wrist and sniffing it, until he nodded '*itr*', then '*aywa*', yes.

I had envisioned some funky little shop where an elderly gentleman speaking French with the soft-rolling Lebanese 'r's would show me rare blends. But when the chauffeur triumphantly opened the door, it was to usher me into the black marble mausoleum of a Saudi perfume company that had a Paris branch I'd never set foot in. I was swiftly attended to by a slender young man who looked as me as tenderly as if he'd wanted to slide a spoonful of mastic ice cream between my lips, and who proceeded to display what seemed to be half the contents of the shop. I'd never had experience with oil-based perfumery, which this was: when they're not airborne by the evaporation of alcohol, the aromatic materials develop in a very different way. The face I was making clearly demonstrated I had not partaken of Botox. Then the piastre dropped: most of these fragrances seemed to be directly inspired by Western blockbusters.

My purveyor of sweetness saw he wasn't reeling me in so he took out choicer bait: the oil of Taif roses, the most highly prized in the Middle East, grown at an altitude of two thousand metres in Saudi Arabia. As I kept shaking my head, he whittled it down to a few drops for one hundred dollars. I started backing away slowly as he held out the little crystal bottle:

'If you want one perfect thing in your house ...' he cooed.

I didn't have the heart to tell him I didn't like rose that much.

'*Ma chérie*, you should have asked him to show you the oud,' said Habibi over his morning coffee after greeting me with a spine-dislocating hug. 'The Saudis keep giving me oud chips. I'll get you some before you leave.'

He never did.

Had I dawdled in the shop a little longer, I would have discovered a bit more about the thriving Middle Eastern perfume industry. The region must have the highest per capita consumption of perfumes in the world. Several companies, based locally or in Europe, put out blends that are either secular formulas, knock-offs of successful Western scents in oil, Western formulas with oud added, or products of a new perfume family called 'French-Oriental' that typically revolves around an oud and rose accord.

The use of oud, also known as agar wood, aloes wood or jinko, goes back centuries: it is mentioned in the Sanskrit Vedas and the Old Testament. It is the olfactory signature of the Middle East, though it is also highly prized in India and Japan, and either burned in chips to steep rooms or clothing with fragrant smoke. The essential oil can be mixed with other notes, often in homemade blends, or worn as a 'backdrop' to Western fragrances.

Oud is a resin that forms when a type of mould infects the heartwood of certain varieties of the *Aquilaria* species, found mainly in Southeast Asia. Though the trees can be injected with the mould, much of the production is harvested in the jungle. Only one out of a hundred trees is infected, but as it's impossible to tell from the outside, many trees are just chopped down for expediency, most often by poachers, endangering the species (it is now protected). And as the merchandise usually goes through several middlemen, it can, at times, become costlier than gold.

The note first made its first official appearance in a mainstream Western fragrance with M7, launched in 2002 by Yves Saint Laurent under Tom Ford's tenure. The scent's animal sexiness was underscored by the first ever perfume advertisement featuring full-frontal male nudity, dangly bits and all. Since then, in their mad scramble to woo the juicy Middle Eastern market while they figure out how to get a peg in China, Western perfume companies have been chucking oud into their more exclusive collections.

With some niche enthusiasts, oud has become a fetish of sorts, to such an extent that I've come to think of it as a *cultural* artefact rather than an aromatic material. Its leathery, medicinal, smoky facets tick all the right boxes among hardcore perfume aficionados: because it is such an acquired taste for Western noses, embracing oud is a way to show you can take it like a man, whatever your gender, not unlike 'skank'.

But it's more than that. I suspect the current Western oud mania is the latest guise of Orientalism. Back when few people actually travelled all the way to the 'Orient', dabbing on Shalimar garbed in one of Poiret's harem get-ups was enough to turn you into an odalisque. Now that booking a flight to Dubai or Bangkok is no more complicated than planning a weekend in Normandy, we need a more genuine marker of exoticism, and oud is that, just as patchouli used to be for the hippies. As such, it is every bit as much of an oriental fantasy as Shalimar; an appropriation of the Other through what is most other: smell. Not to mention that splurging on it puts us, at least symbolically, on the same footing as the sheiks.

But years before the West discovered the Middle Eastern 'authenticity' of oud, the Middle East itself was indulging in its own Western fantasy by hiring French perfumers to produce fragrances in the grand French tradition, just as that tradition was dying out in the West. Thus the house of Amouage, founded in 1983 by a member of the royal family of Oman, kicked off with a classic aldehydic floral – probably the first straight-faced interpretation of the genre since First in 1976 – composed by the veteran Guy Robert, who authored Madame Rochas (1960) and Calèche (1961). Jubilation 25, put out in 2007 to celebrate Amouage's twenty-fifth anniversary, was yet another *fin-de-race* fragrance, this time from the chypre family.

In English, the expression *fin de race* translates as 'degenerate', which doesn't convey the aristocratic connotations of the French. Someone who is *fin de race* is the last of a noble lineage whose excessive inbreeding has exaggerated every family trait to the point of caricature: think Habsburg chin or Bourbon nose. And Jubilation 25 is indeed *fin de race* in many ways. In it, Femme's dark plum has fermented into booziness; its spices have risen to nose-tingling prominence, while oak moss is further darkened by leather-scented clouds of incense. The scent also exhibits more than a trace of Opium's DNA, with its aldehydic shimmer and resinous-fruity myrrh facets: its author Lucas Sieuzac is, after all, the son of Jean-Louis, who co-signed Opium with Raymond Chaillan. Somehow this opulent iteration of a classic genre fatally compromised by the public's disaffection and increasingly stringent regulations on the use of oak moss feels like a chypre that doesn't know that chypre is dead. The news may not have reached Oman yet.

The Middle East may actually be the last place really to *believe* in perfume enough to infuse lifeblood into the decaying flesh of old-school French perfumery, much in the way that the Japanese are the last who truly believe in fashion, as anyone who attends Paris fashion week can observe. As though classic French perfumery, after a century of oriental fantasies, was catching its last reflection in a mirror held up by the real Orient.

* * *

Bertrand has worked for Amouage and his boozy-fruity, incense-laden and spice-sprinkled Jubilation XXV, the masculine pendant of Jubilation 25, was an early foray of his into oud, a note he may have pioneered in the West by introducing it in the 2001 Sequoia for Comme des Garçons. Now he's nose-deep in it: his clients are frantic for French-Oriental. At first I grumbled that the note had become such a cliché I didn't even want to smell what he'd come up with: I was ouded-out. Then I caved in and asked him what it was about oud that excited him so, and watched as he became the intense, enthusiastic man-child his favourite raw materials always seemed to turn him into.

He described oud as resinous, ambery, with smoky and leathery facets reminiscent of castoreum, all of which I could figure out for myself. And then it all poured out: why it spoke to him so deeply. And I understood that oud might well be the missing link between his calling as a perfumer, his far-flung journeys and the antique tribal art he collects ... He told me how oud had a sour milk note that reminded him of the bottom of old wooden South Ethiopian milk pots he'd bought, of temples in the Himalaya, of kitchens in Nepalese houses, of the antique Tibetan furniture he owned which still held the smell of the rancid yak butter that had been kept inside it. Oud *was* the smell of the Other for him, I realized, but in a carnal, primal way rather than as a piece of exotica. It was the smell of the Other *within*; a connection with odours once familiar to our common core of humanity he sought out in his travels.

Oud is burned like incense, and, as it happened, I'd been thinking about incense along those lines. Why it was so profoundly moving to people from so many cultures; why cultures all over the world seemed to burn incense-like substances in their spiritual quest ... As though some part of us remembered the bits of fragrant wood our remotest ancestors might have thrown into their campfires in East Africa where our species was born, or perhaps as they migrated through the Arabian Peninsula, the very places where the trees that yield incense grow.

And going, further back, I thought of smoke. Couldn't the smell of one of the things that made us human, the ability to make fire, be somehow imprinted into our genetic memory? Was that why fragrances with smoky notes had such an emotional impact on us? Bertrand agreed as we quietly breathed in the penetrating wafts of the oud.

'We've cut ourselves off from our ancestral memories,' he said. 'Smells are one of the only ways to regain what we've lost.'

36

Monsieur likes to keep whatever bit of cigar he hasn't smoked in the evening and light it up again in the morning. It is, he says, a form of fidelity: what you loved one day, you love the next. No woman of his has ever found this habit bearable, he adds. Then he corrects himself: none save me.

Yesterday's cigar could be an apt description of our love affair, couldn't it? He keeps setting it down, then lighting it up and drawing it to his lips again. The one thing I can be sure of is that he'll always come back for a puff; that he'll never let it go out completely, though we'll never go anywhere, least of all to Seville.

I wasn't surprised when he told me it wouldn't be possible to make the trip after all. I'd never actually believed it would. I didn't even truly want it to happen, though I might not have had the courage to turn the trip down if he'd come through ... Between us, Seville had better remain what it has always been: the vivid background against which I first emerged in his life. Seville is an aura that trails after me, a reinvention not to be tested against reality, just like Duende. And so the night of the Madrugada has passed once more without me. At any rate, the trip would have been a disaster: for the first time in eighty years, the processions had to be cancelled because of torrential rains.

Today is the date I set for Duende to be finished, one year to the day after my first session with Bertrand. It's not.

We did meet three weeks ago at a café terrace near the Paris Opéra, but only to catch up. I'd drenched myself with Duende just before our appointment. He analysed it as I reported back on my first round of beta-testing in London the previous week. I'd sprayed mod 63 on a group of friends and I couldn't recommend a better ice-breaker for a party: these people hadn't known each other when they walked in but within fifteen minutes they were mingling and chatting as they sniffed each other's wrists. I'd also skin-tested Duende on my twenty-year-old American students at the London College of Fashion, without telling them what it was. They'd instantly broken into blissful grins, and begged to know when it was coming out.

It had been an odd experience to follow Duende on so many different skins after spending a year being the sole tester of its various iterations. It had turned sour on one young woman, and intensely, soapily floral on a few others. One male friend had brought out the blood note quite strikingly. The green, aldehydic structure held out for hours on some, while others rushed to the tobacco-y tonka bean base notes. Of course, my friends couldn't very well say they didn't like it, though one did venture that it wasn't quite edgy enough for his tastes, but my students' spontaneous reactions hinted at a crowd pleaser. Perversely enough, that was a slight cause for concern. My students' tastes *do* tend to run towards the cloyingly sweet: one of the most popular raw materials during my courses is ethyl maltol, which is literally what gives its flavour to candyfloss.

Bertrand nodded: he might decrease the ethyl vanillin, a powerful vanilla note with caramel effects, to tone down the sweetness. He also thought that the floral notes might be a little too heady but, then, I'd practically spray-painted myself and there was so little space available behind that tiny sidewalk terrace table that I was practically sitting in his lap …

In my new capacity as a muse-slash-evaluator, I had my own remarks to make. I felt that the lavender had been sucked up by the flowers, though Bertrand assured me he'd used the same percentage as in the previous mod we'd loved. And I would have liked the base notes to have a little more texture and darkness. Patchouli maybe, he suggested, to give it a little chypre-ish vibe? But no, we shouldn't be adding new notes. No more second-guessing. We were close, almost there … Before parting on the sidewalk in front of the Opera, we set an appointment for 22 April. I knew he wouldn't keep it.

He hasn't. I'm starting to feel we'll never be done, precisely because the scent is almost done: the last adjustments could involve dozens of mods. But, as the French say, the best part of love is climbing the stairs, and in this labour of love it is witnessing the creative process, taking part in it, that is Bertrand's true gift to me. The staircase has just turned out to be steeper and higher than I'd thought.

So I'll just have to wait for Bertrand, and wait for Duende. Anyway, Seville has always been about the waiting. Waiting for the beauty to happen – the *duende*. Waiting for chance to present me with a new adventure. Isn't the Spanish/Arabic word for orange blossom, *azahar*, thought to spring from the very word meaning chance, *azar*? Chance is what led Román to fall into my footsteps that day on the Puente de Triana, when he asked me where I was going and whether he could walk with me. I took that chance without knowing that, some day, the night he gave to me would blossom into a fragrance …

37

What do you wear to the French perfume industry's biggest event of the year? Killer cleavage, that's what. More skin to spritz with Femme in the original formula, just to remind yourself and everyone around you of what perfume used to be and can still be when left half a chance. After all, when asked what they consider to be *the* perfume, Femme is what half the perfumers answer – the others say Mitsouko. Pity the industry isn't living up to that. The top awards have just been nabbed by two of the blandest launches of the year.

This evening, the Fragrance Foundation France is handing out the equivalent of the Oscars crossed with the People's Choice Awards (the public gets to vote online), and let's just say that, if anyone sneezes in the sumptuous Belle Époque ballroom of the Grand Hotel Intercontinental, the industry will grind to a standstill. I'm here because, for the second year running, I sat on the jury of the Specialists Award, given out by journalists, evaluators and bloggers to niche products. My fellow jurors and I have known the result for a week but, when the winner is announced, one of them smirks:

'Do you realize we've just given out the award for a niche perfume to a *celebrity* fragrance?'

Unlike the American Fragrance Foundation, which has a specific category for celebrity perfumes, the French have none because there are practically no celebrity perfumes in France. What we have are *égéries*, the actors and models who front advertising campaigns. The word comes from the name of the Latin nymph Egeria: according to the legend, she secretly advised her lover, King Numa Pompilius, on matters of state religion. In English, it translates as 'muse'. But although the septuagenarian Jean-Paul Guerlain proclaims that all his feminine scents were inspired by women, there are no muses in the perfume industry, at least none that we know of. What Loulou de la Falaise was to Yves Saint Laurent, Gala to Salvador Dalí, Kiki de Montparnasse to Man Ray, Marlene Dietrich to Josef von Sternberg or Gena Rowlands to John Cassavetes? That's never been bottled.

There are, however, a handful of European products that do hint at a muse-perfumer relationship. They fit into the celebrity fragrance slot, but barely, because they're so quirky and original you've just got to believe they reflect their namesakes' decidedly un-Photoshopped personalities – unlike American celebrity juices that give the feeling the stars just slapped their names on the label to rake in the royalties (one of the rare exceptions being Sarah Jessica Parker, a perfume aficionada who had very definite ideas about Lovely). Christopher Brosius' Cumming for the actor Alan Cumming is 'a scent that is all about Sex, Scotch, Cigars and Scotland' – the sex part somehow involving rubber and leather. Miller Harris's L'Air de Rien for the singer and actor Jane Birkin, a fierce moss and musk blend that ranks high indeed on the skank-o-metre was based, Birkin told *Vogue*, on 'a little of my brother's hair, my father's pipe, floor polish, an empty chest of drawers, old forgotten houses'. Rossy de Palma, the striking *belle laide* who graced Pedro Almodovar's cult *Women on the Edge of a Nervous Breakdown*, asked for a rose that would also speak of earth, spices, volcanoes and Africa. For good measure, Antoine Lie and Antoine Maisondieu threw in a drop of blood to conjure the rose's thorns in her Eau de Protection for État Libre d'Orange.

The brand went on to produce a second celebrity scent with the cult gay artist Tom of Finland, but it was their third offering in that line that whipped up the perfect scented storm. For the occasion, the house forewent its provocatively erotic pitches, names and visuals: the scent was called Like This after a poem by the Sufi mystic Rumi. With it, perfumery's bad-boy brand seemed to be coming back into the fold of serious, artistic niche houses, so all was forgiven by perfume lovers, who embraced it enthusiastically. The composition itself, a burnished, burning essay in tones of orange, was arrestingly original but as cosy as a weather-worn tweed jacket: beautiful enough to stand judgement on its own, which is how it was evaluated by the jury. But it certainly didn't hurt that it had a red-hot muse to up its cool factor: Tilda Swinton, Oscar winner, fearless actor, style icon and unearthly beauty.

As Mathilde Bijaoui steps onstage to accept the Specialists Award for Like This, I wonder whether, once the stardust is swept away, her creative partnership with Tilda Swinton couldn't be the closest thing to mine with Bertrand: a hybrid project halfway between bespoke and commercial perfumery, predicated on an individual's story but conceived for public release. I'm curious to find out just how much of herself the actor really put into its development so, after congratulating Mathilde, I ask her the question. She confirms: she really did develop Like This with and for Tilda Swinton.

'I conceived it for her, first and foremost. I wanted it to touch her.'

But the term 'muse' gives her pause.

'I don't know that I'd call her that. To me, a muse is someone you're in love with …'

A co-author, then? Mathilde nods.

'Yes. Definitely.'

A few days later, Mathilde Bijaoui and I resume our conversation in the sun-drenched offices of Mane on the verdant Île de la Jatte, once a bucolic playground for the Impressionists, now one of the poshest suburbs of Paris.

A slender, striking beauty with her tangle of dark curls, Mathilde owes her vocation to her father, a keen amateur chef who took her to food markets to make her smell things. When she visited ISIPCA on an open day at the age of thirteen, she couldn't believe there was actually such a profession as perfumer, and immediately set out to get the scientific *baccalauréat* she needed to be admitted to the school. She wasn't yet thirty when she was singled out by Étienne de Swardt, the owner of État Libre d'Orange, to make proposals for Tilda Swinton's scent.

The star's gender-bending persona would have dovetailed neatly into the house's iconoclastic stance but she wanted to take another direction. 'Scent means place to me: place and state of mind – even state of grace. Certainly state of ease,' she would later write for the press release. 'My favourite smells are the smells of home, the experience of the reliable recognizable after the exotic adventure: the regular – natural – turn of the seasons, simplicity and softness after the duck and dive of definition in the wide, wide world.'

Swinton, who had been wearing Penhaligon's Bluebell for so many years she no longer smelled it, was no perfume aficionada, though she quoted the very classic Joy, Calèche and 24 Faubourg as scents she was fond of. More tellingly, she forwarded a list of the smells that most touched her: simple and delicate flowers like sweet peas and honeysuckle, but also lapsang souchong tea, single malt whisky, wood smoke, bonfires and fireworks on Guy Fawkes Day, which happened to be her birthday.

Smoke appeared clearly as a leitmotif, a boon for a perfumer working on a carte blanche brief. A first session introducing Tilda to raw materials confirmed this: she was strongly drawn to immortelle, an oddball note with its burnt, foody facets of curry and maple syrup. Perhaps its name rang as a subconscious call from Orlando, Virginia Woolf's immortal male-to-female hero and one of Swinton's landmark roles in Sally Potter's 1992 film adaptation of the novel? She never said.

Immortelle made perfect sense to Mathilde Bijaoui, whose synaesthesia makes her see the smell in orange. So orange became another leitmotif. It was the colour of Tilda Swinton's hair; of the dress her character wore in Luca Guadagnino's *I am Love*, which she was shooting at the time. It was even in the name of the brand putting out the fragrance. The idea was so impeccably consistent, but so simple, that Mathilde wondered whether it wasn't just plain simple-minded, but it worked, and she went on adding layers of orange. Over the course of their conversations, Tilda had brought up the fact that she was a ginger, so ginger went in. Ginger called for pumpkin, and a pumpkin accord came about to soften the blend. It worked well with Tilda's request for something homey; Mathilde envisioned a kitchen where a pumpkin pie was baking. Then she added carrot and mandarin: more orange, more flavours. Again, this echoed what Tilda was experiencing as an actor at the time: in *I am Love*, her character is a Russian-born Milanese trophy wife who falls in love with a young chef: 'You have no idea of the match I'm doing between sniffing here and tasting there,' she told Étienne de Swardt. Using food notes may have also been a way for Mathilde, who discovered the olfactory world with her amateur-chef father, to sneak something of her own biography into a scent where she'd been given such a free hand ... After all, she and the actor, born Katherine Matilda Swinton, share practically the same first name: another reflection in the mirror.

Swinton followed the development process closely over a dozen sessions, never missing an appointment and playing the game with utmost seriousness. She appropriated Mathilde's proposal so fully she once had to call her to get a fresh batch of one of the mods: she'd worn it all. She was also the one to find the name, taken from the title of her favourite poem. She suggested it to Mathilde and Étienne when they came to pick her up at the train station. 'That's also when she said that the scent made her think of sex in the afternoon, in her own bed,' Mathilde recalls. Developing Like This, she says, was a happy experience: no hitches, no glitches.

'It's very different from the way we usually work. She trusted me. She never poked her nose into the formula. I'd rather that than someone who tries to understand everything and understands nothing … It was real perfumery.'

If I'd been looking for a reflection of my own experience developing Duende as seen through a perfumer's eyes, I'm not sure I've found it. Clearly Tilda Swinton, while eminently sensitive to smells and possessed of impeccable taste, was not enough of a perfume nut to delve into the technical details. The dynamics might also be different between two women. I need a second opinion, and I know just where to get it.

The socialite, heiress and fashion muse Daphne Guinness has stated in interviews that not only did she make up her own perfume blends, but that she used to extract the smell of tuberoses herself in a DIY version of enfleurage: 'I would collect them and put them all on greaseproof paper with a kind of gel, and then you leave it for a few days. Then you'd scrape off the gel and have a sort of essence …' Therefore, she is clearly one of us, only more chic, richer and more famous: enough to commandeer one of the edgiest perfume brands in the world and the talent of one of Givaudan's top perfumers.

With his mess of blond corkscrew curls and leather jacket, Antoine Lie could be cast as a cool sexy streetwise cop in an existentialist French *détective* movie. Though he seems much too mild-mannered to play tough, I suspect he's got the kind of quiet authority that wouldn't require raising one's voice. Nevertheless, when I meet him in Givaudan's sleekly designed offices near the Arc de Triomphe, it is with the vague hope that the spectacular Ms Guinness has turned out to be more of a handful for 'her' perfumer than I've ever been with 'mine'. Antoine admits he wondered what he was in for. After Comme des Garçon's Christian Astuguevieille had offered him the job, he'd Googled Daphne and, judging from her flamboyant sartorial options, had expected a forceful, eccentric lady.

'But in fact she was quite subdued. She was entering a world she didn't know at all. She wanted to learn about it, and she was very respectful of the perfumer's work.'

Daphne had already sent over the two 80s scents she had blended to create her own signature fragrance, one based on tuberose, the other on patchouli. She wanted something along those lines, but which would be uniquely hers. Antoine liked the idea of working on a single person's specifications, just like perfumers used to do in the era of classic perfumery; of composing a bespoke perfume that would be commercialized. He couldn't envision himself doing it for just one person, but with something that was meant to come out, he felt he'd be more in control. What's more, Daphne's templates appealed to him: he started out as a perfumer in the 80s and he relished the perspective of revisiting the rich materials and sensuous notes of the era, a welcome change from the panel-tested products he worked on for the mainstream.

Just as Mathilde with Tilda, their first session was focused on raw materials, to find out which ones Daphne preferred. Patchouli and tuberose: confirmed. She didn't like spices much, but amber and oud conjured memories of her trips to the Orient. Incense she fell in love with during that session, he believes, though Daphne has spoken in interviews of her memories of High Church Masses. But, unlike Tilda, Daphne didn't talk much about her life during that meeting.

'She spoke about her olfactory memories, but mostly she focused on the materials,' explains Antoine.

From what was a very clear initial proposal, he set out to build a fragrance based on overdoses of the main notes, treated as blocks that could be perceived throughout the development rather than as a fluidly evolving scent with small facets, a construction Daphne agreed to. The result is a lush, broad-stroked descendent of 80s floral powerhouses like Poison. I suggest this to Antoine, who hadn't thought of it but doesn't deny it: if Poison was, he suggests, 'Fracas meets Shalimar', Daphne could be thought of as 'Fracas goes to India'. I also pick up a rich-bohemian-at-the-beach, tanning-lotion vibe – a result of tuberose's coconut facets combined with salicylates and amber. Antoine explains – that quite matches Daphne's childhood memories of her hippie-chic summers at Cadaqués.

But my hopes of finding a more persnickety muse than myself are dashed: the development took all of two face-to-face appointments, two or three waves of mods and the addition of a bitter orange top note to liven up the blend before Daphne-the-muse was happy with Daphne-the-perfume.

'She figured if Christian had given me the job,' says Antoine, 'then she could trust me.'

So much for finding a mirror that can catch the reflection of a muse to meld with mine … Tilda Swinton and Daphne Guinness may have infused their perfumes with their personal tastes, followed their development, inspired the perfumers they worked with to go where they wouldn't have gone otherwise, but their stories aren't my story. And it's not because I'm not famous. Perfume is immediate and intimate; it is blind to limelight. In confronting themselves with it, they had to shed their public personas, so that, whatever they put of themselves in those bottles, it goes beyond image. That is why we can be touched by Like This or Daphne, make them ours, even without knowing the women who lent their image to them. But also why, however many questions I ask, I'll never know the truth of the story behind each, the unspoken secrets that were told through scent. At least, not beyond the word that struck me. It came up both times.

The word is trust.

38

It may have been a mistake to leave me alone for so long with Duende 63. It's been over two months since our last session and now N°63 *is* Duende for me, despite its technical flaws. But Bertrand finds it too harsh and dry, so he's been toning down the incense and spices and reconsidering our decision to leave out musk: he feels he needs it to wrap the floral note, smooth it out, give it more amplitude.

But every time I pick out a blotter without looking at the numbers, it's to find I've fallen back on N°63: I can't wrap my nose around the new mods. Perhaps I'm a little flustered by the set-up. Today, we are not in the lab but in the adjacent office so that Bertrand's trainee Pascale can sit in. A pharmacist from Marseilles who decided to change careers and attended a perfumery school in Grasse, she is a sweet, considerate person and I feel comfortable with her. But the fact that this work session is not one-to-one gives it a different tone, more focused and technical. It is a normal part of the process as the term of our project draws nearer, but one that is making me feel a little dispossessed. Duende is becoming a product.

Bertrand's also lobbed a curve ball at me in the form of N°72, a variation on N°63 to which he's added clary sage and ambroxan, a woody-ambery material present in many masculine fragrances, to boost the tobacco note.

'It's not Habanita,' he explains, 'but it's the cigarette the guy is smoking as he feels you up in the crowd. I want to keep that note. It goes with the story.'

Pascale says strangers stopped her to ask her what she was wearing the day she skin-tested N°72, though she'd sprayed it on hours earlier. Bertrand adds he tried it out on skin as well and that it's got the volume and persistence that are missing from N°63.

I'm torn. I can tell he is very interested in this new direction and I'm tempted to go along with it because, after all, the man knows what he's doing … but no. Much as I enjoy the tobacco top note, I find the ambroxan too masculine. Besides, it swallows up what I love most about the opening of N°63, that gorgeous ethereal green whoosh that feels so exhilarating when I first apply it. If the idea had come earlier in the game, I might have gone with it, I tell them. At this point, it strays too far from the options we've taken. As Frédéric Malle once told me, you can't kiss every pretty girl you see in the street.

After much discussion and comparative sniffing, we settle on N°90, which seems the best balanced and most radiant of the new mods. I spray it on one arm and N°63 on the other: as I hold them out to Bertrand and Pascale I must look like the statue of Christ in Rio de Janeiro … N°90 is better rounded, smoother and more diffusive than N°63, and I do find it delicious, even undeniably gorgeous, but I feel it doesn't have quite as much bite.

Bertrand decides to lay both formulas flat; he'll copy them down side by side so he can see where he's at and what's been lost along the way, in order to readjust dosages and reintegrate materials he's dropped. Pascale dictates the concentration and quantities, which are expressed in parts per ten thousand; the materials are grouped both according to their effect (citrus, green, floral) and volatility (as top, heart and base notes). This is the first time I see the formula in full detail. Oddly, instead of dispelling the romance of perfume-making, hearing this long, austere list is a disconcertingly emotional experience: as each material appears in turn, the formula conjures the whole history of Duende, as though it were a coded transcription of a year in my life. I remember the day they came up, the mood we were in, the things we told each other ...

Beeswax? I mentioned it right after I'd expressed doubts about the first two proposals, when I told Bertrand about the kids who collected the wax from the penitents' candles to make balls. That day I took him back with me to Seville with my words; he said he felt he'd lived that experience, maybe in another life.

Iso-amyl salicylate? The blood note Bertrand introduced last May after I'd given him Lorca's book on *duende*.

Algix and glycolierral? Sitting at a café terrace before the August holidays, a little bored by the names of all those chemicals, and being asked by Bertrand if I knew what I wanted.

Magnolan? Flint and floral. This came up in September just before I went to visit Olivier Maure at Art et Parfum near Grasse, and I feel a flutter in the stomach at the idea that I'll be going back when the first industrial batch of Duende is weighed.

Tagetes, angelica and cassis? They appeared a year ago in May and made a comeback in October, the day I was cross with Bertrand for being away so much and not listening to me, and I brought in Habanita and N°5, and the scent took a new direction, and almost came to a screeching halt.

African Stone? I made him smell it the day I told him that, in Monsieur's opinion, the scent wasn't erotic enough. It was dropped between 63 and 90, though not on purpose – Bertrand just forgot about it. He'll put it in again, but he'll also experiment with civet.

Vanilla? The Moan.

Black and pink pepper? They appeared just as I came back from the land of the perfume ban. Who knows, maybe they were a wink from my pepper-addicted father.

Luisieri lavender? The first time I felt true emotion when I smelled Duende, because lavender reminded me of the beginnings of my affair with Monsieur, but also because Duende went from being a project to being a perfume that day.

I'm also meeting the new players. There are two types of synthetic musk: ambrettolide because it's got the effects of ambrette, a lovely natural material that contains vegetal musk but also smells of Poire William liqueur, and therefore reinforces the rum in the top notes as well as the waxy effects. Globalide for its tobacco and amber facets. He also added styrax to reinforce the waxy, smoky and balsamic 'candle and chapel' notes.

Duende is becoming a fairly complex product with its forty-three materials. Each of the groups plays a role in the story. Citrus, green and floral notes for the orange tree. Lavender for the smell of the colognes in the crowd. A tobacco effect for their cigarettes and Habanita. Incense and its boosters, like pink and black pepper, along with beeswax, for the religious procession. And the oriental base with vanilla, benzoin, styrax and tonka bean, again for Habanita.

Bertrand combines the two formulas to get back some of the bite and vibrancy of N°63 while keeping the lushness of N°90. But the new formula won't be weighed immediately. I'll have to wait until next week to smell the result. Bertrand suggests grabbing a cup of coffee next door: it's been a fairly intense session, he needs a break and we haven't had a chance to catch up in a month.

I miss the long, rambling conversations we had when we started meeting. Since that time, his career has taken a great leap forward. Now he's so busy he's wondering whether he won't have to turn his one-man operation into a proper company. But it's not just that. His status is shifting. 'Bertrand Duchaufour' is becoming a brand, and that brand is starting to get top billing, above his clients'. It might not yet have made him a household name, but that may be changing too. When we met in that radio studio, it was his first live media appearance. Now he's just taken part in two prime-time television travel pieces. This type of exposure is a fairly new phenomenon: up to now, only the in-house perfumers of luxury brands had crossed over into the mainstream media.

My own position within the perfume world has evolved as well over this past year. With Duende, I've taken the leap from the virtual world into the lab. Nothing will ever be more enthralling than the first step I've taken through the looking glass, but now that I've taken it, I want to go further. Bertrand has taught me so much: he's trained me to become … what? Étienne de Swardt of État Libre d'Orange calls himself a *parolier du parfum*, a 'perfume lyricist'. I could go along with that.

Bertrand shakes his head:

'No. You're not a perfume lyricist. You're a *critic*.'

Granted. But then, in the sense the writers at the *Cahiers du Cinéma* were critics in the 50s.

If perfumery is to be compared to an art form, I suggest to Bertrand, I've been thinking it should be cinema. Not because they create similar objects but because both have evolved along similar lines. Modern perfumery and movies were born around the same time, in the late 19th century. Both quickly burgeoned into an industry that aimed to draw in the public with appealing, widely available products, in contrast to the other arts, which were just then moving into the avant-garde and its assaults on aesthetic conventions. Both industries developed a system in which the creative forces were studio staff, compelled to express themselves within commercial constraints. And both found keen observers from outside the seraglio. For Hollywood, it happened in France when all the American movies produced during World War II finally made it across the Atlantic at the same time after the war. A group of passionate young critics discovered that directors like Alfred Hitchcock, Billy Wilder, Douglas Sirk or

Nicholas Ray had managed to develop and express their personal style and formal language despite working within the strict constraints of Hollywood: they called their stance the *politique des auteurs*, 'the politics of the author'.

Perfumery took fifty years to catch up and produce its equivalent of the *politique des auteurs*; it was certainly harder going since, unlike directors, perfumers only started getting credited over the last decade. But today it has one, of sorts. Just like the *Cahiers*, perfume critics have singled out *auteurs*, Bertrand among them, tracked their production from one house to another, and analysed their signature style. And although our work together will not turn me into a perfumer in the way that Godard, Truffaut or Rohmer went from writing about movies to directing them – I have no intention of becoming one and I'm nowhere near comparing myself to these great filmmakers – it does open new possibilities. To paraphrase Jean-Luc Godard, 'art and the theory of art, beauty and the secret of beauty', perfumery and the explanation of perfumery, belong to the same continuum.

Whatever the label – 'lyricist', critic, writer – I am a perfume lover trying to think through perfume, both in the sense of thinking the matter through, and in the sense that perfume is one of the languages I use to understand the world. In this new realm I am exploring, a vanilla pod can turn into a cigar and a cigar can grow into a bale of hay; the bale can spit out an almond and the almond turn to poison. The magic is conjured by connecting smells to words. Technical knowledge does not dispel that magic. In fact, it's just the opposite. The more I learn, the stranger and more magical it gets. This is what perfume is teaching me: that once it is unmoored from the product, the process of thinking through perfume need never stop.

39

'Come on … concentrate! You're hopeless … Hey, will you just *concentrate?*'

Bertrand, Pascale and I are sitting at a round table, looking for all the world as though we are gearing up for a poker game. Bertrand's been dealing the blotters. He's teasing me because I keep getting them mixed up or dropping them as I try to put them in the right order and spread them out in a fan.

'That's because I don't play cards. I hate losing too much.'

'Listen to her … She hates losing. What'll we hear next?' he chuckles.

I give him a little kick.

'Hey, I may be hopeless with blotters, but as soon as it's hot enough, I'll show you I know how to handle a *real* fan. It's not for sissies.'

Maybe he's had a really nice weekend, maybe he's happy with the way Duende is going or maybe he's just happy to see me. For whatever reason, Bertrand is in a particularly cheerful mood today. So am I, for that matter. It's only been a week since our last session, and it's exhilarating to be moving forward so quickly again. I'm not even cross at him for being half an hour late. It gave me the chance to have a chat with Pascale. It was the first time I could discuss my work with Bertrand with someone who is part of the process. Pascale is the one who's been weighing the formulas of the successive mods for the past six months; she's lived behind the scene and seen how Bertrand deals with other projects and clients ... Since my conversations with Mathilde Bijaoui and Antoine Lie, who both said how important it had been for them to have been trusted by their 'muses', I've been worried about being too much of a pain for Bertrand. About not trusting him enough; about sticking my nose too far into what was essentially his business. But Pascale said we were doing just fine as far as she could tell. I knew what I wanted, was all. She hadn't heard any complaints. I could have hugged her then; and again when she told me total strangers had been asking her what she was wearing when she tested N°90. So I was in a pretty buoyant mood myself when Bertrand showed up, boasting he'd swum two kilometres that morning 'with his fingers in his nose' – the French equivalent of 'without breaking a sweat', but it conjured a particularly vivid picture of Le Nez flapping his elbows in the municipal swimming pool ...

I've finally managed to spread out my blotter fan to compare N°90 to mods 96 to 101. In some of the new mods, the waxy effect is more sharply defined in the top notes: Bertrand has introduced aldehyde C12 lauric, which will also help with the diffusive power. And, as promised, he's tried out two different animal materials, African Stone, which brings out the woody incense notes, and civet, which is smoother and rounder.

Duende has evolved into a strange, unclassifiable creature Bertrand calls an 'oriental cologne', which sounds like an olfactory oxymoron. But in fact, it does seem to cut through the full spectrum of fragrance families: fougère, green floral, white floral, oriental, woody incense ... Just about everything but chypre.

'It's even got masculine, aromatic effects! That doesn't bother you, I hope?' jokes Bertrand.

'Certainly not. But what'll Michael Edwards say?'

'He'd call it a floriental.'

Michael Edwards is the author of *Fragrances of the World*, a classification designed to assist marketing and sales staff, based on a 'fragrance wheel' comprising fourteen different families defined by a set of dominant notes. The 'floral oriental' family, 'soft, spicy orange flower [that] melds with piquant aldehydes and sweet spices' on an ambery base, is descended from the 1905 L'Origan; it includes L'Heure Bleue, Bal à Versailles and Poison.

'Mind you, it's a floriental because I *wanted* it to be a floriental!'

I stare at him blankly. Meaning?

Bertrand is pulling a typically Duchaufourian face at me: the cocky, slightly defiant look of the mischievous kid who's fooled the grown-ups.

'I've taken the easy road by putting in vanilla. I wanted to give it twice as many chances of being successful, of living on ... So ... *Allez, hop!* I orientalized it.'

True, I *have* been after him to take it easy on the musk and vanilla; I *am* concerned that the combo will nudge Duende too close to some of Bertrand's other compositions, and make it a little too commercial if the balance isn't just right. But I've got to acknowledge that we did need the musk to make the floral note hold longer: I've been testing N°90 on skin and it's definitely an improvement. Still, I don't see why he's cocking his head on his shoulder, looking as though he's about to stick chewing gum into my hair.

'Why are you looking at me like that? Come on, spit it out!'

'Well, vanilla wasn't necessarily part of the story.'

Tsk, tsk ... Once more, Mr D. is ever so slightly rewriting history.

'Oh, please! It was *totally* part of the story from the moment I brought in Habanita!'

'True. You're right', he nods.

'Come on, *concentrate!*' I tease him.

While Pascale is weighing a couple of new mods, we go down to the café next door, pick a table in the sun and order cheese-burgers – they'll be vile, but the health-food lunch bar doesn't have a terrace – still excitedly discussing Duende. I tell Bertrand that while wearing N°90, images of gardens and flowers kept flitting through my mind; memories of sucking nectar out of flowers ...

'Exactly! Nectar! That's very important. Nectar makes you think of honey, which makes you think of beeswax – bees adore orange blossom.'

He explains that he works with words as much as he does with smells. Words echo other words, colours, and colours turn into odours, situations, places ...

'... like that plaza full of orange trees in full blossom buzzing with drunken bees. From there you go to beeswax candles, from candles to incense.'

It all comes back to what we were discussing almost a year ago on this very terrace: the 'remote and accurate' connections that underlie poetic images. How you intuitively grasp the connec-tions between the smells in a story to create a form, even before becoming aware of all those connections. How, once you become aware of them, you develop the consistency between the olfac-tory and the narrative, and strengthen the connections so that they are subconsciously perceptible to the wearers.

I remember the email I sent to Bertrand after he'd almost given up on Duende: I'd told him that the orange blossom and incense accord was good *because it existed in reality*. It's not a matter of copying reality, but of deciphering the secret harmo-nies of smells that have existed together for mankind over the centuries. How long has incense been burned in countries where orange trees grow? How long have beeswax candles been lit at the same time as incense? This is what we were telling each other about oud, incense and smoke the other day: that they connect us with smells we've lost and sometimes don't even remember losing. Perfumes rouse those unconscious, age-old memories. They connect the remotest past to our deepest soul, our soul to the body, and our body to the world. What are those connections between the smells of our bodies and, say, the fattiness of aldehydes or beeswax, the lactones and indolic notes in white flowers, the animalic whiff of civet or African Stone, if not ways of reaffirming the continuum between ourselves and the animal, vegetable and mineral kingdoms?

Modern civilization has ripped us out of that continuum: perfumers, if their work is true and free, restore us to it. But to do so, they must seek out that truth in their own stories: for Duende, the beautiful memories of church incense that allowed Bertrand to transcend the hardships of a strict Catholic upbringing. When I told him my story, those memories connected with mine: with the voluptuousness of Catholic rites transcended into the glory of a spring night in the south of Spain. What is most intimate is what will speak to others. Perfumers build the labyrinth in which we lose ourselves out of all those secret harmonies and connections. They bring out its beauty: reinvent it so that it can be felt by all. 'The poetic act', the French poet Mallarmé wrote at the turn of the century, 'consists of suddenly seeing that an idea splits up into a number of equal motifs and of grouping them; they rhyme.'

Now that we've found the rhymes, the thread that would guide us into the labyrinth and out, I ask Bertrand whether this is how he envisioned the fragrance when we first set out.

'Well, yes, pretty much. What's incredible is that it's all in there. All the elements we started out with. We just learned to put the puzzle together.'

We pick at our cheeseburgers for a while in silence.

'Incredible,' Bertrand finally lets out. *'C'est ça, un parfum.'*

That's what a perfume is ...

So we talk about other things as we finish our lunch; of the way you have to come to terms with the past to move into the future, and first it's about our lives but then it leads us back to what we do for a living. Bertrand tells me he's reconsidering things he'd rejected earlier on in his career. He's starting to break free, he says, working on 'crazy accords' he hopes his clients will accept; making his move into new territories.

That's when I make mine. I hadn't meant to bring it up today, but all of a sudden it seems like the right moment to ask him whether he'd be game to work with me again on another perfume, if the right project came along.

He would. This time, I don't faint.

40

'You'll see. I've changed *everything!*'
I stop dead in my tracks.

'What do you mean, you've changed everything?'

I've just bumped into Bertrand at the biennial raw materials exhibition organized by the Société Française des Parfumeurs. This is the very event where we'd re-connected after that fateful radio show, and attending it has been a striking way of measuring just how far I've come in a world where I took my first steps just three years ago. The last time, I drifted among unfamiliar faces. Now I keep running into the lovely people who've taught me, guided me, spent hours discussing the art of perfumery with me: Isabelle Doyen, Sandrine Videault, Mathilde Laurent, Dominique Ropion, Mathilde Bijaoui, Élisabeth de Feydeau, Annick Le Guérer, Pamela Roberts, Olivier Maure … It feels as though the cast of my fragrant Wonderland has come together for the grand finale. But I hadn't expected to see Bertrand, who's popped in one hour before closing time and who is now dragging me from booth to booth to smell raw materials.

Our latest exchange was a bit stormy. I'd written to him to express my doubts about the latest mods he'd done: I felt the larger quantities of vanilla and musk and the addition of rose were pulling the scent towards too-familiar grounds. He replied that he'd already moved forward, based on a mod picked by L'Artisan Parfumeur in the meantime. That riled me up so much my finger-tips were tingling. I shot back that, if my opinion wasn't consid-ered relevant, there was no longer any point in my testing.

It wasn't the fact that L'Artisan Parfumeur had weighed in on the development that irked me: that was entirely normal. After all, much as I considered Duende to be my baby, it was their product, to be part of their collection. They weren't going to let Bertrand run off for a year and a half and come back with a finished product, however much they trusted him. But I was annoyed that Bertrand hadn't kept me informed, and hadn't mentioned which mod they'd picked. What if the submission they'd preferred was precisely the one I felt was the least interest-ing? There wasn't much I could do if they had.

Fortunately, minutes later, he was apologizing for the way he'd put it: actually, the people from L'Artisan Parfumeur had felt the same way I did about his most recent tweaks, and he saw their point, so he'd gone back to mod N°90 and was going to move forward from there. Clearly, I didn't have to worry about the brand's aesthetic options. I shouldn't have done in the first place, considering Bertrand's body of work with them.

Still, changing *everything*? Bertrand, what have you done to our baby?

But he's got such a huge grin on his face, and seems so happy with his new take on the formula, that there's only one thing for it. Take the leap of faith.

'You're scaring me … But I trust you. Absolutely.'

Once I'm back in the lab one week later, I'm so distracted at the prospect of finding Duende radically altered that I can't smell properly. I'm pecking my nose at my blotter fan like a hen that's afraid it's lost its chicks, frantically asking about this or that material rather than analysing the new mods.

Bertrand reassures me: everything is still there, though in different doses, except the jasmine absolute because it felt too cloying. The formula was so tightly packed the accords couldn't breathe freely, he explains. As he was working on another product, he found a new way of doing the orange blossom accord that was more expressive, more diffusive and less costly. More expensive doesn't necessarily mean better, and in this particular case he is convinced that this formula for the orange blossom accord is an improvement. I agree. It is more faceted: brighter, greener, more cologne-like in the top notes and more sensuous in the base notes.

When Bertrand asked me, almost one year ago, whether I knew what I wanted, I couldn't say. I just knew that, if the orange tree was there, there was no one under it. We've spent months wandering in the labyrinth, trying to summon the presence, the *soul* that is conjured when a heartfelt story finds its fullest expression in scent. When we almost got lost because I'd made us take a new turn with Habanita, and Bertrand was bumping into dead ends, I'd written to him: 'You'll get there, and your perfume will be heartbreakingly beautiful.' And then I went to light a candle to Mary Magdalene.

She's just answered my prayer.

Now I know what I want. I want *this*: mod N°123. Is there any magical thinking involved? Today is the 23rd. I was born on the 23rd. One-two-three: two people and an orange tree.

It's so obvious there isn't even a decision to make. Everything is there, but everything is *clearer*. Bertrand has spilled sunshine into the Sevillian night. It isn't my olfactory memory of Seville I am regaining when I breathe in Duende N°123: it is the emotion of walking into beauty. The *duende*.

So, is that it then? Are we done?

Bertrand shakes his head.

'When you find the accord, it's easy to make it evolve up to its near-final form, which corresponds to ninety-five per cent of the completed formula. But you've got to realize one thing: the main effort of the perfumer – and, I believe, of most artists when they are working on a piece – bears on the last five per cent.'

This fine-tuning, he explains, is the most crucial part of the process. For a fragrance to be successful, there are two prerequisites: a good hook, those expansive top notes that will draw you in, seduce you. And a well-balanced formula, which will give it the maximum volume it can reach.

'To give it soul, you'll work for ages until it's perfectly polished. No. Not perfectly polished. Perfectly within the idea and balance you want to give it.'

How long will that take? I've more or less resigned myself to picking up the work in September – the French summer holiday debacle is soon approaching.

And then, a few days later, an email drops into my inbox. The final version will be selected in two weeks by Sarah Rotheram, the CEO of L'Artisan Parfumeur.

This is it.

Before this last session, Bertrand and I are meeting again, this time with Alissa Sullivan, who is in charge of olfactory development for the brand and studied at ISIPCA.

Though I'd exchanged a few emails with Alissa when we discussed potential names for the scent, we only met a few weeks ago at the press launch of a new product. I spoke with her and Nick Steward, the head of marketing and product development, about getting together to share our impressions on the last steps of the development. And here we are.

The last mods Alissa evaluated pre-date the latest shake-up, so she's come to find out where the formula is at, and to weigh in on possible adjustments to make before Sarah comes in with Nick to take the final decision.

I've been skin-testing Duende N°123 and getting extremely positive comments. The orange blossom accord is marvellously expansive during the first couple of hours and very long-lasting.

But it slacks off a bit in the heart notes, and Bertrand has been experimenting with ways to give it even more volume.

This is where things get tricky. We both love the way N°123 smells, and that's pretty much where we'd like to keep the scent. But there is no magic ingredient that can amplify the volume of a perfume without skewing its form. You've got to alter proportions in materials that are already in the formula or add something new, but whatever you do it won't smell exactly the same.

Bertrand has come up with three new mods, all based on N°123, all beautiful, and different enough to present clear options.

N°125 is the brightest. Alissa feels it might be a good pick for a spring launch, though we all feel that Duende is rich and complex enough to be a perfume for all seasons.

N°127 has an even stronger orange blossom note, but the dose of Luisieri lavender has been reinforced to bolster the incense accord. As a result, it is the darkest, most balsamic and most sensuous of the three, but it's also a little bit flatter in the top notes.

As for N°126, Alissa and I are both less keen on it. It isn't so much an orange blossom because of a new material, alpha-damascone, one of the components of rose. Bertrand is nothing if not stubborn: this is the third or fourth time he tries working in a rose note.

'I'm insisting wickedly, because rose always amplifies the volume of the heart notes.'

'Yes, but it also changes the smell,' I point out.

'OK, so I've moved the cursor slightly, but bear in mind that you have to take in the whole of the line on which the cursor is moving. The note is a little less orange blossom in N°126, but isn't it worth toning it down to achieve more volume?'

Bertrand insists we try 126 on skin, so as both our wrists are scented with 125 and 127, we squirt the crooks of our elbows and indulge in a session of perfume-testing yoga, six arms twisting around like a fragrant Shiva as we sniff each other out.

'It's girlier, younger,' Alissa comments, but not as though these were necessarily positive attributes: she's just stating the facts.

'I'm fed up with young-and-girly,' I grumble.

'Hey, wait a minute, this isn't girly,' Bertrand protests. 'It's not an innocuous perfume.'

'But it *is* sweeter,' I insist.

'You're absolutely right. The damascone fruits it up, it juices it up. It makes the heart notes redder. Like red apple juice. Whether that's necessary or not, I don't know.'

Alissa doesn't seem to think it is. Like me, she finds that the new note pulls the scent towards a more common, more commercial register and, again, she doesn't mean it in a positive way.

'That's the problem with rose,' I argue. 'Because it *does* give more volume and appeal to a floral note, it's a trick of the trade, which means adding it makes the perfume a little less original.'

'You're right,' says Bertrand.

Alissa shakes her head.

'I'd rather we didn't go for something too sweet. The last two perfumes we've launched are both sweet and fruity …'

As I let out a sigh of relief, Bertrand nods.

'All right. You'll have the final choice anyway.'

Meanwhile, the darker mod N°127 is asserting itself on our skins. Alissa asks whether it wouldn't be possible to do another mod based on it, but with more contrast in the top notes so that they're as fresh as those of N°125.

While Bertrand tinkers with his formula, Alissa and I go on smelling the mods and chatting. This is the first real chance I get to have a proper talk about Duende with someone from L'Artisan Parfumeur. As it turns out, they started following the project quite early on, before Bertrand even mentioned to me that they were interested. In fact, Sarah, Nick and Bertrand had been discussing developing a fragrance based on a novel just about at the time I came into the picture. So when I told him my story, which I had always intended to use in a novel about my adventures in Seville (I may have even mentioned that to him at some point), it all clicked. This doesn't make the way the stars aligned at that particular moment any less serendipitous; on the contrary: it could have been anyone but me, any story but mine.

But like Bertrand hearing the call of his materials, perhaps I heard his call for a story – heard it in his willingness to be carried off by my tales; in his yearning for faraway places. Heard it in Al Oudh, the first fragrance of his I understood from the inside, which meant we could talk. Heard it in Nuit de Tubéreuse, which I felt had been meant for me, though it hadn't. Heard it in Vanille Absolument: a call from Habanita though it took me months to realize it.

So it turns out that Bertrand wasn't the only one to hear *me*. Thinking back, I guess there was never another option than L'Artisan Parfumeur for Duende: they were the ones who pioneered the idea of perfume as travel sketchbook, and they've always based their scents on stories rather than sticking on a story after the fact ... Though as I smell the three current versions of Duende on my arms, I tell myself that what started out as a narrative scent has also become an incredibly complex *abstract* perfume, as well as a reflection on the history of perfumery with its references to cologne, fougères, white florals and orientals ...

It may also be, quite simply, one of the sexiest scents L'Artisan Parfumeur has ever put out although, oddly enough, it isn't flamboyantly feminine. And that too is what I'd envisioned it to be: a fragrance that would reflect the story of a man and a woman; the original one I told Bertrand, but also our own creative journey.

* * *

Bertrand re-emerges from his lab shaking a 5ml phial and, from the look on his face, Alissa and I can tell he's pretty happy with the result. In fact, as we lean in to sniff fresh blotters of 125, 127 and the new 128, he's practically whinnying with excitement.

'N°128 is a little less settled. But ... whoo! My first impression is *very* good. This is going to be tops!'

'Of course, with the parents it has, it can only be good,' I gloat.

Bertrand throws a sly glance at Alissa.

'I've added a product that will explode the budget.'

She looks up from her blotters.

'Rose ... but this time, in the form of rose oil,' Bertrand explains.

This isn't stubbornness; it's sheer pig-headedness. But it works. We all find N°128 more contrasted and expansive than N°127.

'And even more mysterious,' Bertrand adds.

He's sprayed his forearm with N°128. I lean forward to smell it on him. His hairs tickle my nose and I back off, stifling a sneeze.

'I've put in body hair absolute!' he jokes.

I rub my nose as the three of us burst out laughing.

'Perfumer hair headspace? Now *that* would be a new concept!'

As Bertrand pops back into his lab to prepare another set of samples for Alissa to test over the weekend before they are presented to Sarah and Nick, he's still hooting about how good he thinks Duende has become.

I've never seen him so exhilarated. Does he feel that way every time he reaches the end of a project? For his sake and for the sake of the art of perfumery, I hope so. But I'd rather think none of his perfumes has ever been so lovingly nurtured by its muse.

And now I'm about to present our brainchild to the lady who'll unleash it into the world. I'm sure she'll love it.

41

D uende has a name now, the name by which it will go out
into the world.

'Séville à l'aube' came to me as I thought back to the first time
I visited Bertrand in his lab. That day I told him about Seville
and he gave me a sample of a tuberose scent he called Nuit de
Tubéreuse, without knowing what role the criminal, carnal
tuberose – that corruptress of a flower – had played in my life.
Segueing from the night of the tuberose into Seville at dawn feels
like a continuation of the story that began that day. The story
that is ending now.

Seville was always about waiting for the beauty to happen.
But it was also about cutting – cutting losses, cutting loose.
'Seville at dawn' is right: it sets time back into motion. Dawn
was when Seville dug into my flesh like a knife and, as I run my
fingers over the scars I've reopened so many times since, it is my
story I am reading in Braille.

Román and I had made it through the night, giddy with incense and wine, bruised tender from being pressed against one another by the crowds, raw-lipped from dawn kisses rough with dark bristles, voices raspy with laughter and smoke, the pearly mist of exhaustion veiling our eyes. We'd made it through the night, pinned to the orange tree – green leaves, white flowers, golden fruit against the lavender sky – and it felt like a victory to have lived until dawn; until the crowds trailed slowly towards the looming purple Giralda swallowing the gilded floats and left us standing alone.

We stopped at a stall to buy floury hot chocolates and the fried strings of dough called churros, *and sat on a bench under an olive tree. Román held out a burning piece of dough for me to bite. I licked the caster sugar off his fingers and drank the chocolate, thick and dark as congealed blood. A giggling cross-dresser tottered on platform shoes, followed by a gaggle of gypsy boys good-naturedly chanting 'Guapa, guapa y guapa' to her as they'd done to the Virgin of La Esperanza minutes before, and we laughed too.*

Then it was just Román and me, his arm around my waist, our steps ever slower as our giddiness waned and our limbs grew numb, and we drew from the well of the night one last time, knowing what we had lost in defeating it. Soon the sun would tear the last shred of darkness off our skins; the sun was another world and, for a few minutes, while dawn still embraced us, we hovered at the edge of goodbye – it is at dawn that the duende *murmurs its dark sounds most poignantly and we were straining to hear its song.*

Román left me in the doorway of the Hotel Simon. I stumbled through the hotel patio and made my way up the staircase tiled with copper and blue azulejos *to reach my tiny rooftop room ... He'd promised he'd come back after Easter, so I waited.*

All the orange trees burst into blossom on Easter morning and, under her bridal crown, Seville traded her black lace mantilla for a blonde one. At three in the afternoon, the voice of the city changed into a deep hum, as though thousands of nectar-maddened bees were converging towards the honeycombs of the bullring. Even the light was different: it had been waiting for the first bullfight of the year to take on its purest colours, a gold-drenched blue streaked with swallows darting over the ochre sand – white arcades, ink-black bulls, sepia blood, tiny glittering doll-men waving fuchsia and buttercup, then poppy-red capes ...

I'd planned to hop onto the train the next day and spend an afternoon in the Prado before returning to Paris. I forgot to leave. Instead, I drifted into the limbo between Holy Week and the April Fair, caught in the immutable present tense of Seville's dusty walls – blind men roaring 'Para hoyyyyy!' as they touted their lottery tickets, one-armed bandits wailing their electronic siren song in cafés strewn with sawdust, bits of waxed paper and gamba shells, the white alleys where every path led back to the statue of Don Juan raising a paper glass filled with rainwater to the old men who huddled on the tiled benches of the Gardens of Murillo ...

I was waiting for Román. I knew nothing of him, except that I amused, excited, seduced him. He was no more to me, perhaps, than my effect on him, and the effect Seville was having on me. So I waited for him to keep the promise Seville had made me. It's not a gift if you have to ask for it: I asked for nothing. I waited. The beauty of Seville was a gift in itself, and I wanted to lose myself in it.

I left a forwarding address at the Hotel Simon when I went back to Paris after the April fair. I got his letter in September.

You are as unpredictable and fickle as I am. You left Seville without telling me, but no matter: whatever happens, we are forever bound by our magical night.

But now, I need to see your smile, your pink lips, your honey eyes, your porcelain skin wrapped in black lace. I remember them perfectly well even though you left me no memento. Send me a poem, a picture, a dried flower, the imprint of your lips on a piece of paper ...

Instead of a flower, I sent myself. But when I arrived in Seville in the first cool nights of fall, she turned her back on me with Carmen's mocking laugh. Román was going for a week to the beach. Perhaps we'd meet up when he returned if I was still around ... We'd have a drink somewhere.

I didn't see him that time, but I came back often, not for Román but for Seville. And I did sometimes see him when he hadn't gone to the seaside or the sierra. We spent some nights together, never more than one at a time. I never knew when we met whether he'd take me home. But I was filled with so many stories back then that I was never truly with him; besides, our nights could only be faint echoes of the Madrugada. Yet Román was always watching over me as though Seville had entrusted me to him, a stranger who never truly understood the mores of the city and wandered into his arms to rest for a few hours from her adventures ...

Then I met the Tomcat and there was no more Seville. And then, after the orange blossom had beckoned in Marrakech, Seville was where I went to decide whether I'd leave my marriage. The Tomcat said he wouldn't be there when I came back. I hoped he meant it.

It had been ten years since my last trip. I learned through one of Román's friends that he'd sold his loft in the city and gone to live in a village in the Sierra Norte. The friend gave me his phone number. I called. He said to come. I thought, if he asked me to stay, I'd never go back to Paris.

I got up at dawn to take the coach and fought sleep throughout the three-hour ride so I wouldn't miss my stop. When I came out of the coach I caught my foot on a step and tumbled into Román's arms as he staggered back laughing, engulfing me in his warm man-smell. As we bumped along country roads in his dusty Jeep, he told me about the rare flamenco songs he was recording for the state archives. We stopped and got out to walk. He told me the Spanish names of birds and herbs, and we sat on a worn stone bench, and I wondered at my blindness: I'd never seen this beautiful man for what he was. When I told him, he laughed: he'd always seen me for what I was, he said, but he didn't say what.

His house in the Sierra was almost empty. A cold-water shower in the courtyard. A white bed under a mosquito net. A threadbare faded velvet couch. Books. That afternoon I read Lorca in Spanish lying on the couch, my feet in Román's lap. Then as we prepared dinner in his tiny tiled kitchen which opened onto a courtyard where chickens were pecking, I sang Cole Porter's 'Night and Day' and Román stopped chopping tomatoes and tapped the beat on the counter with his knife and his fingertips until the jazz was tinged with flamenco and we were almost dancing.

It was too hot to eat much, so mostly I drank the wine, and then he said:

'You know, we won't sleep together tonight, guapa.'

I rose from the couch unsteadily, bumped my head on a beam and burst into tears at the sudden sharp pain. He laughed and hugged me, kissing my forehead, and rocked me until I started laughing too.

'Why not?'

'We had our magic night, guapa. It won't get any better.'

'It could be different.'

'Can you see yourself living here in the Sierra? You belong in Paris. We're not in the same book, Carmen, la del Canada ...'

Oh, but yes we are ... You can *noli me tangere* me all you want, my love, but you can't make love to a writer without her stealing your story to turn it into hers, just as I've stolen Seville so that it could be turned into the air I breathe.

'Román' was the name we agreed on back then if I should ever write our story. But the real Román, the one who has another name that doesn't mean 'novel' in French, might read me with his lips moving a bit as he makes out the words in a language not his, and it will be as though he were kissing my ghost. Perhaps he'll reach out to me, not to take me away from my life but to wrap me again in his arms for a moment, because there will always be the love that burned itself out in a night and vanished at dawn to become another kind of love tempered in the crucible of the magic night. And there will always be the night in the Sierra when I trusted him with the knife to cut me free, so that Seville could become the capital of memory from which I drew the gold of words; as immaterial as perfume, and as sweet.

Perhaps the woman who shares Román's life will wear my Séville à l'aube, and it will be as though our magic night were wafting back into his dreams. That's all right: Séville à l'aube will no longer then be mine. If I catch a whiff of it as I'm walking past a stranger, I'll smile at her, but she won't know why. By then, I may be trailing in the wake of another love, the better to lose myself again in Seville when I come back to it …

But, for now, Séville à l'aube is still my Duende: an exquisitely meandering journey with the best of all possible companions – the intense, endearing, infuriating, headstrong, brave, generous and brilliantly gifted perfumer who invited me into his invisible world and opened himself up to mine. That the result of that journey would be a thing of beauty I've never doubted. Duende was already beautiful when it was barely a sketch, and each of the hundred shapes it took on as it grew and shifted was striking. It was heart-rending to let go of them: they suggested new alleys in the labyrinth, leading to other landscapes. But we had to listen to what Duende was asking us to become.

One single perfume cannot express all that perfume can say, all that a perfumer can say, all that a perfume lover can hope for. It can only be the most beautiful expression of the spirit that inspired it at a given time and place: as singular, flawed and moving as the people who brought it into the world, and as the ephemeral world they created together. Duende is all the choices that were made and all the choices that weren't. Séville à l'aube is the result of those choices; not a conclusion but a story to be reinvented by all those it touches.

Meanwhile, I've got an appointment with Mr D. He says he's got something new to show me. I've never lived a longer dawn.

Acknowledgements

I couldn't be more grateful to Bertrand Duchaufour for his trust, time, creativity, generosity and friendship, and for making me the most beautiful gift a perfume lover could ever hope for.

A million thanks to Jenny Heller, my marvellous editor at HarperCollins, for sparking the idea for this book as well as for her invaluable support and advice. And thank you so much to Helena Nicholls, who took over the project with such grace and enthusiasm, as well as to Georgina Atsiaris, Elen Jones and all the team at HarperCollins, and to Susan Opie for her keen eye.

I have a huge debt of gratitude to my wonderful agent Homa Rastegar at A. P. Watt, and to our friend Alexander Greene, who set everything in motion. Thank you also to all the team at A. P. Watt, especially to Linda Shaughnessy and Louise Lamont.

I am deeply grateful to L'Artisan Parfumeur for launching Séville à l'aube into the world: thank you to Sarah Rotheram, Nick Steward, Alissa Sullivan for making it happen, and to Florian Pedemanaud for being so supportive.

I am also deeply indebted to my friend Octavian Coifan for sharing his encyclopaedic knowledge so generously and for rereading the historic parts of the manuscript (any remaining errors are mine), as well as for the countless hours we spent discussing and smelling things.

Thank you to Olivier Maure and Michel Roudnitska, who invited me into the magical realm of Sainte-Blanche, and to Sandrine Videault for instilling some of its spirit into me from afar.

My heartfelt thanks to everyone in the perfume industry who gave me their time and shared their knowledge and stories with me: Christian Astuguevieille, Abdes Salam Attar, Mathilde Bijaoui, Élisabeth de Feydeau, Étienne de Swardt, Pierre Dinand, Isabelle Doyen, Marie Dumont, Vero Kern, Mathilde Laurent, Annick Le Guérer, Antoine Lie, Serge Lutens, Frédéric Malle, Annick Menardo, Pamela Roberts, Dominique Ropion, Patrick Saint-Yves …

This book would never have been possible without the vibrant online perfume community where I found so many friends and sources of inspiration: all my gratitude to my fellow bloggers, with special thanks to March Dodge for allowing me to quote extensively from her Perfume Posse posts, to Nathan Branch, who provided a crucial bit of advice at the right moment, and a big shout-out to all of the readers of Grain de Musc who contributed so many thoughtful comments to my blog.

Thank you to my London support team, Laurent Delaye, Clare Bradley, Basia Szkutnicka, Kit Davis, Nicola Stephens and Silvia Pezzali; to Guy Scarpetta and Rachel Laurent for the house at the back of the garden where I wrote several chapters, and to Catherine Greslot, who made sure I was decently fed while I was shackled to my keyboard.

I also tenderly dedicate this book to the memory of Fabiana Leonor Heifetz, who gave me more love, advice and support during the brief months of our friendship than most people could in a lifetime.

And finally, *un beso* to Enrique, wherever he is. He knows the real story.

Bibliography

Baudelaire, Charles, 'Head of Hair', *The Flowers of Evil*, transl.
  William Aggeler (Fresno: Academy Library Guild, 1954)
Burr, Chandler, *The Perfect Scent* (New York: Picador, 2007)
Calkin, Robert R. and Jellinek, J. Stephan, *Perfumery: Practice
  and Principles* (Hoboken: John Wiley & Sons, 1994)
Clark, Kenneth, *The Nude: A Study in Ideal Form* (Princeton:
  Princeton University Press, 1953)
Coifan, Octavian, *1000 Fragrances* (1000fragrances.blogspot.
  com)
Corbin, Alain, *The Foul and the Fragrant: Odour and the French
  Social Imagination*, transl. M. Kochan (Oxford: Berg
  Publishers, 1986)
DeJean, Joan, *The Essence of Style* (New York: Free Press, 2005)
Détienne, Marcel and Vernant, Jean-Pierre, *Les Jardins
  d'Adonis: La Mythologie des parfums et des aromates en Grèce*
  (Paris: Gallimard, 2007)
Diderot, Denis, *Rameau's Nephew*, transl. Ian C. Johnston
  (http://records.viu.ca/~johnstoi, 1762)
Edwards, Michael, *Perfume Legends: French Feminine Fragrances*
  (Sydney: Michael Edwards & Co., 1996)

Ellena, Jean-Claude, *Journal of a Perfumer* (London: Penguin, 2012)

Ellena, Jean-Claude, *Le Parfum* (Paris: Presses Universitaires de France, 2007)

de Feydeau, Élisabeth, *A Scented Palace: The Secret History of Marie-Antoinette's Perfumer* (London: I. B. Tauris, 2006)

Flaubert, Gustave, *Correspondance, Tome 1* (Paris: Gallimard, 1973)

García Lorca, Federico, *Theory and Play of the Duende*, transl. A. S. Kline (www.poetryintranslation.com, 1994)

Genet, Jean, *Un Captif amoureux* (Paris: Gallimard, 1986)

Genet, Jean, *Journal du voleur* (Paris: Gallimard, 1949)

Gilbert, Avery, *What the Nose Knows: The Science of Scent in Everyday Life* (New York: Crown Publishers, 2008)

Greer, Germaine, *The Female Eunuch* (London: Paladin, 1970)

Haskins, Susan, *Mary Magdalen: Myth and Metaphor* (New York: Riverhead Books, 1993)

Holley, André, *Éloge de l'odorat* (Paris: Odile Jacob, 1999)

Jaquet, Chantal, *Philosophie de l'odorat* (Paris: Presses Universitaires de France, 2010)

Jellinek, Paul, *The Psychological Basis of Perfumery*, 4th edition (London: Chapman & Hall, 1997)

Kaufman, William I., *Perfume* (New York: Dutton, 1974)

Le Guérer, Annick, *Le Parfum: Des Origines à nos jours* (Paris: Odile Jacob, 2005)

Le Guérer, Annick, *Scent: The Mysterious and Essential Power of Smell* (London: Chatto & Windus, 1992)

Piesse, Septimus, *The Art of Perfumery* (Philadelphia: Lindsay and Blakiston, 1857)

Pliny the Elder, *Natural History*, transl. John Bostock (London: G. Bell and Sons, 1892)

Proust, Marcel, *Swann's Way*, transl. C. K. Scott Moncrieff (Forgottenbooks.org, 1913/2008)

Reverdy, Pierre, 'L'Image' in *Nord-Sud N°13* (Paris, 1918)

Rilke, Rainer Maria, *Rodin & Other Prose Pieces*, transl.
  G. Craig Houston (London: Quartet Books, 1986)
Robert, Guy, *Les Sens du parfum* (Paris: Osman Eyrolles, 2000)
Roudnitska, Edmond, *Le Parfum* (Paris: Presses Universitaires
  de France, 1989)
Roudnitska, Edmond, *Une Vie au service du parfum* (Paris:
  Thérèse Vian Éditions, 1991)
Schiaparelli, Elsa, *Shocking Life* (London: V&A Publications,
  2007)
Stamelman, Richard, *Perfume: Joy, Obsession, Scandal, Sin* (New
  York: Rizzoli, 2006)
Turin, Luca, *Parfums: Le Guide* (Paris: Hermé, 1994)
Turin, Luca and Sanchez, Tania, *Perfumes: The Guide* (London:
  Viking, 2008)

Index